W. Boehm and H. Prautzsch

Numerical Methods

Wolfgang Boehm and Hartmut Prautzsch

Numerical Methods

vieweg

CRC Press
Taylor & Francis Group
Boca Raton London New York

CRC Press is an imprint of the
Taylor & Francis Group, an informa business
AN A K PETERS BOOK

Mathematics Subject Classification (1991): 65-01, 65 B XX, 65 C XX, 65 D XX, 65 F XX, 65 G XX
CR Subject Classification (1991): G .1, G.2.2, I.3.5

First published 1993 by Vieweg and A K Peters Ltd.

Published 2018 by CRC Press
Taylor & Francis Group
6000 Broken Sound Parkway NW, Suite 300
Boca Raton, FL 33487-2742

ISBN-13: 978-1-56881-020-1 (pbk)
ISBN-13: 978-1-138-41317-7 (hbk)

Preface

This book addresses undergraduate students in mathematics, computer science, and all other fields of science and engineering. Readers who have studied advanced calculus and have a basic knowledge of linear algebra should be well prepared for the material.

The book provides an elementary introduction to methods of numerical analysis. It exposes fundamental algorithms for the solutions of widely varying mathematical problems. The book prepares the reader to deal with related questions and to apply principles introduced here to new problems.

To encourage the programming of the methods, explicit algorithms are given throughout. For better understanding the algorithms are presented in a pseudo code. Simple examples further illustrate the methods.

This book has grown out of a one-semester upper-level undergraduate course repeatedly held by the first named author at the Technical University of Braunschweig. Based on a book also by the first author on the same subject the second named author taught courses in Numerical Computing at Rensselaer Polytechnic Institute. These experiences led to the writing of the present book.

Both authors are most grateful to Carl de Boor for his careful reading of the entire manuscript, his many helpful comments and suggestions for improvements. Special thanks also to Dr. Joachim G. Niebuhr for preparing the TEX files that were finally used in publishing this book.

Troy, N.Y., 1992

Wolfgang Boehm
Hartmut Prautzsch

Table of Contents

I Fundamental Concepts

An **algorithm** is an unambiguous finite rule applicable to a class of **problems**. After finitely many steps, an algorithm transforms a problem into another one or provides its solution.

1 Algorithms and Error Propagation

Numerical Analysis deals with the development and analysis of algorithms for the solution of numerical problems. Not every algorithm determines the solution with sufficient accuracy. Because of round-off errors, problems may even not be solvable on a digital computer.

1.1 Algorithms

In Numerical Computing an **algorithm** is a precisely stated sequence of **elementary operations**. It takes finitely many steps to generate an **output** for any admissible **input**. In general, input and output are elements of $\mathbb{R}^n$ and $\mathbb{R}^m$ respectively. Elementary operations can, for instance, include the four basic arithmetic operations, but also the definite integration or the solving of systems of linear equations.

In order to structure an algorithm clearly every algorithm is divided into two parts: Input and output are defined in the first part, the **declaration part**. The elementary operations to be performed appear in the second part, the **instruction part**. They are numbered and if necessary commented. The calculation of a **scalar product**

$$s = a_1 b_1 + \cdots + a_n b_n$$

shall serve as an example. The corresponding algorithm is given below:

Scalar Product

	Input:	$a_1, \ldots, a_n;\ b_1, \ldots, b_n$
	Output:	$s = a_1 b_1 + \cdots + a_n b_n$
1	Set $s := 0$.	
2	For $i = 1, 2, \ldots, n$	
3	form $s := s + a_i b_i$.	

The symbol ":=" represents **substitution**. The instruction **2** "For $i = 1, 2, \ldots, n$" means that the instruction **3** "form $s := s + a_i b_i$" is to be executed sequentially for $i = 1, 2, \ldots, n$. Thus, instruction **3** is subordinate to the (loop) instruction **2**, and this is indicated by **indenting** this line. Initially the variable s equals zero, then s is an intermediate result and finally it contains the solution of the problem. Every algorithm is named and is marked by a frame.

1.2 The Implementation of Algorithms

When an algorithm is performed on a computer or manually, errors occur because concrete calculations are done in a finite subset of the real numbers generally not closed under the elementary operations. Typical computers use just the real numbers

$$\tilde{x} = m \cdot 10^a$$

with an integer **characteristic** a, $|a| \leq q$, and a **mantissa** m, $|m| < 1$, which can be represented as a p-digit decimal number. The numbers $\tilde{x}$ are called **machine numbers** or **floating-point numbers**. The mantissa is said to be **normalized** if $|m| \geq 0.1$. Customary limitations for the characteristic and the mantissa are $q = 99$ and p between 8 and 12. A real number x is represented by a machine number closest to it. This method of representation is referred to as **rounding**.

A computer multiplies two reals x and y by first rounding the numbers to $\tilde{x}$ and $\tilde{y}$, then it determines $z = \tilde{x} \cdot \tilde{y}$, and finally (if $|z|$ is less than 10^q) rounds z to $\tilde{z}$. In general one has therefore $\tilde{z} \neq \widetilde{x \cdot y}$. The other basic arithmetical operations are performed analogously. Together they form the so called floating point arithmetic. Note that neither the law of associativity nor the law of distributivity apply to this kind of arithmetic.

Assuming normalized mantissae, the **relative error** ρ can be estimated as follows

$$\rho := |\frac{x - \tilde{x}}{x}| \leq \frac{0.5 \cdot 10^{-p}}{0.1} = 5 \cdot 10^{-p}.$$

The number $eps := 5 \cdot 10^{-p}$ is called **machine accuracy**. It is also the number of the least absolute value for which $1 + eps \neq 1$ in machine arithmetic.

1.3 Judging an Algorithm

An algorithm defines a map from the set of the admissible input data, a subset of $\mathbb{R}^n$, into the set contained in $\mathbb{R}^m$ of all possible output; i.e. the solutions of the problems. Round-off errors and further errors in the input lead to errors in the output. One requires that these errors be small. A problem is said to be **well-conditioned** if a relative change of the input by *eps* alters the output only slightly. This deviation is called the **unavoidable error.**

An algorithm is **numerically stable** if the distortion of the final result due to round-off errors during the computation is not much larger than the unavoidable error. While a problem may be well-conditioned, not every algorithm for its solution is necessarily numerically stable.

1.4 Notes and Exercises

1. The product $z = x \cdot y$ is well-conditioned.
2. The sum $z = x + y$ is well-conditioned for $xy \geq 0$.
3. The sum $z = x + y$ is ill-conditioned in floating-point arithmetic for $x \approx -y$ and $x \neq 0 \neq y$. This is known as **cancellation.**
4. The system of linear equations

$$\begin{aligned} ax + y &= 1, \\ x + ay &= 0, \end{aligned}$$

   is ill-conditioned for $|a| \approx 1$.
5. The term

$$z = x - \sqrt{x^2 + y}, \qquad x \gg y > 0,$$

   is well-conditioned. If the relative error in calculating the square root equals *eps* then

$$z := x - \sqrt{x^2 + y}$$

   is not a numerically stable algorithm, but

$$z := \frac{-y}{x + \sqrt{x^2 + y}}$$

   is. Why?

2 Matrices

The problems of linear algebra take on a particularly clear and illustrative form if described by matrices. The same holds true for the algorithms that solve these problems.

2.1 Notations

The $n \cdot m$ coefficients $a_{i,k}$ of the **inhomogeneous system of linear equations**

$$\begin{aligned} a_{1,1}x_1 + a_{1,2}x_2 \quad \ldots \quad + a_{1,m}x_m &= a_1 \\ \vdots \qquad\quad \vdots \qquad\quad \vdots \qquad\quad \vdots \qquad & \quad \vdots \\ a_{n,1}x_1 + a_{n,2}x_2 \quad \ldots \quad + a_{n,m}x_m &= a_n \end{aligned} \tag{1}$$

form an $n \times m$-**matrix**

$$A := \begin{bmatrix} a_{1,1} & a_{1,2} & \ldots & a_{1,m} \\ \vdots & \vdots & & \vdots \\ a_{n,1} & a_{n,2} & \ldots & a_{n,m} \end{bmatrix} = [a_{i,k}].$$

The **entry** $a_{i,k}$ of A is located at the intersection of **row** i with **column** k. It is often useful to depict an $n \times m$-matrix as a rectangular **block**:

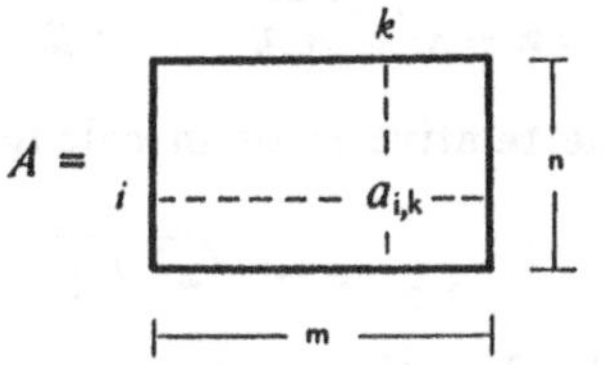

If $m = n$, then the matrix is said to be **square**.

The $m \times n$-matrix $A^t := [a_{k,i}]$ whose rows are formed by the elements of the columns of the $n \times m$-matrix A is called the **transpose** of A:

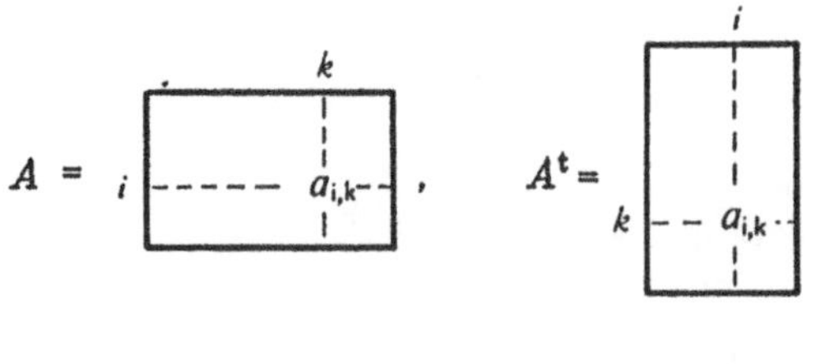

The matrix is **symmetric** if $A^t = A$, i.e. if $a_{i,k} = a_{k,i}$ and $m = n$. The square matrix $I = [\delta_{i,k}]$, with $\delta_{i,i} = 1$ and $\delta_{i,k} = 0$ for $i \neq k$, is called **identity matrix**.

A square matrix U with $u_{i,k} = 0$ for $i > k$ is termed **upper - (right) - triangular**, a square matrix L with $l_{i,k} = 0$ for $i < k$ **lower - (left) - triangular**:

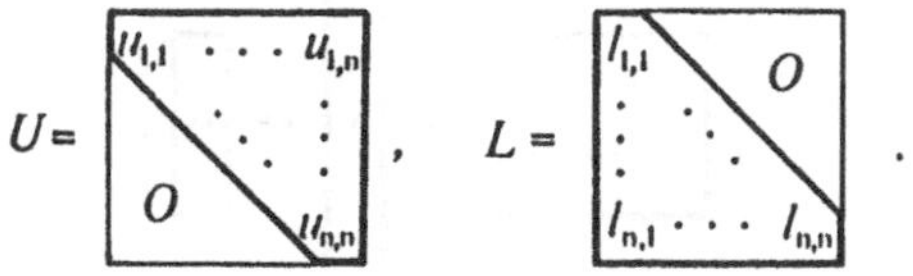

An $n \times 1$-matrix $\boldsymbol{a}$ is referred to as an n-**column** or **column-vector**, an $1 \times m$-matrix $\boldsymbol{b}^t$ as an m-**row** or **row-vector**, and an 1×1-**matrix** c is called **scalar**.

A matrix O with only zero entries is called **zero matrix**, a column $\mathbf{o}$ with only zero entries **zero column**, and $\mathbf{o}^t$ **zero row**.

Finally, a matrix is said to be **sparse** if only a few entries are non-zero.

2.2 Products of Matrices

The **product** $A \cdot B$ of an $n \times l$-matrix $A = [a_{i,j}]$ and an $l \times m$-matrix $B = [b_{j,k}]$ is an $n \times m$-matrix $C = [c_{i,k}]$ where

$$c_{i,k} := \sum_{j=1}^{l} a_{i,j} b_{j,k} \, .$$

This can be visualized by blocks:

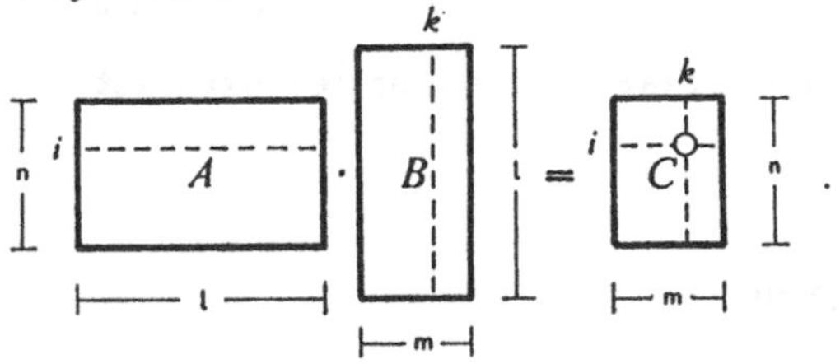

The element $c_{i,k}$ equals the scalar product sum of the entries in row i of A and the entries in the column k of B, short: row i × column k.

A square matrix A with

$$A^t A = I$$

is called **orthogonal** or more precisely **orthonormal**.

Example 1: The numbers a_i on the right side of the linear system (1) form an $n \times 1$-matrix or n-column $\boldsymbol{a}$. Similarly, the variables x_k can be viewed as the entries of an $m \times 1$-matrix $\boldsymbol{x}$. With these abbreviations, (1) takes on a clear and comprehensive form:

$$A\boldsymbol{x} = \boldsymbol{a} \, .$$

The column and row "lengths" can immediately be seen from the block representation:

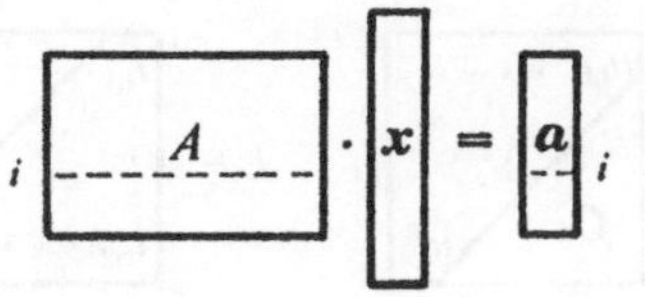

Example 2: The product of two lower triangular matrix is again a lower triangular matrix. This can be checked immediately from the blocks:

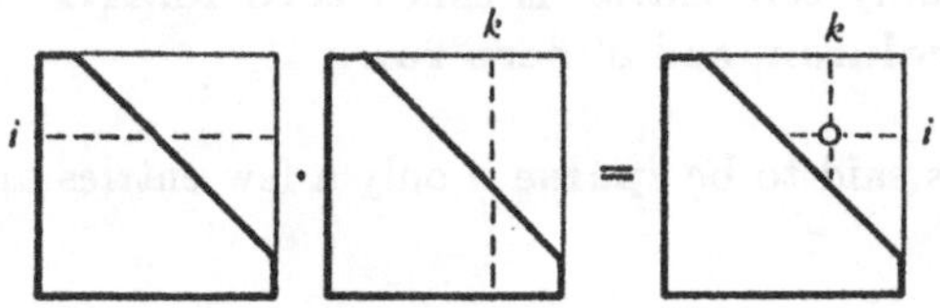

the entry i, k of the product vanishes for $i < k$.

Example 3: The scalar product sum of two n-columns $\boldsymbol{x}$ and $\boldsymbol{y}$, or of an n-column $\boldsymbol{x}$ and a scalar a etc., can be written as a matrix product as well:

$$s = \boldsymbol{x}^{\mathrm{t}}\boldsymbol{y} = \boldsymbol{y}^{\mathrm{t}}\boldsymbol{x} \qquad\qquad \boldsymbol{z} = \boldsymbol{x}a$$

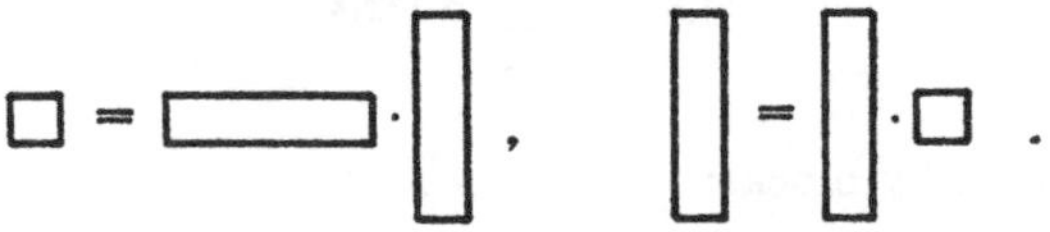

s is a scalar called the **scalar product** or **dot product** of $\boldsymbol{x}$ and $\boldsymbol{y}$.

2.3 Falk's Scheme

The computation of the product $A \cdot B$ of two matrices A and B by hand is aided by an arrangement due to S. Falk. The rows of A are written to the left of the rows of $A \cdot B$ and the columns of B above the columns of $A \cdot B$:

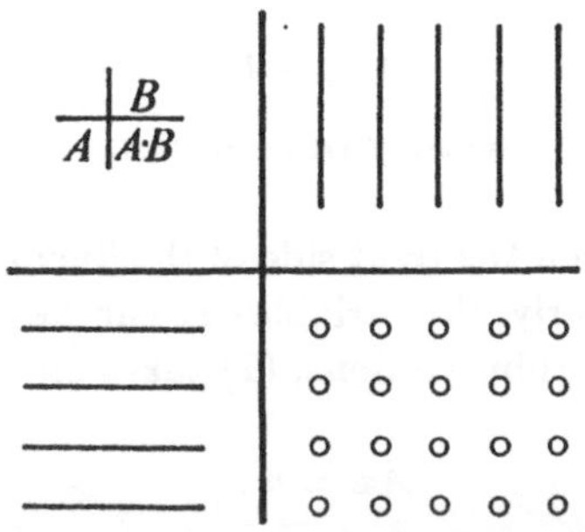

The element $c_{i,k}$ of the product $C = AB$ equals the dot product of the ith row of A and the kth column of B. The symbols of the involved matrices to be multiplied are recorded in the empty upper left square.

Example 4:

	B	1	2	3	4
A	$A \cdot B$	2	1	0	1
1	2	5	4	3	6
2	1	4	5	6	9
0	3	6	3	0	3.

For instance, $c_{2,3}$ of AB is obtained from the scalar sum $2 \cdot 3 + 1 \cdot 0 = 6$.

Example 5: In Falk's scheme, the linear system (1) for the column $\boldsymbol{x}$ of unknowns has the form:

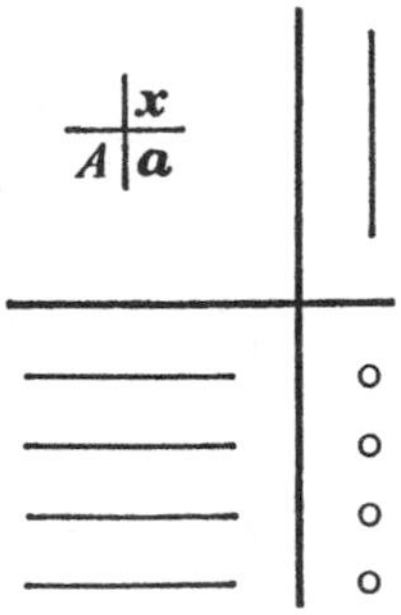

2.4 Rank and Determinant

The m columns of an $n \times m$-matrix A are **linearly independent** if every non-trivial linear combination of the columns does not equal the zero-column, i.e. if

$$A\boldsymbol{x} \neq \mathbf{o} \quad \text{holds for all} \quad \boldsymbol{x} \neq \mathbf{o}.$$

The maximal number of linearly independent columns of A is called the **rank** of A and denoted by rank A. It equals also the maximal number of linearly independent rows of A. The symbol det A signifies the **determinant** of a square $n \times n$-matrix. If rank $A = n$, then $\det A \neq 0$ and vice versa; A is called **non-singular** in this case. Every non-singular matrix A has a unique **inverse**. It is the non-singular matrix A^{-1} which satisfies

$$A^{-1}A = AA^{-1} = I.$$

2.5 Norm and Convergence

A real-valued function $\|\cdot\|$ on the set of all $n \times m$-matrices (n, m fixed) satisfying the three conditions

I. $\|A\| > 0$ for $A \neq O$ and
$\|A\| = 0$ only for $A = O$,

II. $\|cA\| = |c|\ \|A\|$ for all reals c,

III. $\|A + B\| \leq \|A\| + \|B\|$,

and if $m = n$ also the additional condition

IV. $\|AB\| \leq \|A\|\ \|B\|$

is called a **matrix norm**. Condition IV. means that the norm is compatible with the matrix multiplication. The norm is a generalization of the length of a vector. The following are some choices for the norm of an $n \times m$-matrix:

Frobenius norm $$\|A\|_0 := \sqrt{\sum_{i=1}^{n} \sum_{k=1}^{m} a_{i,k}^2}\,,$$

column sum norm $$\|A\|_1 := \max_k \sum_{i=1}^{n} |a_{i,k}|\,,$$

spectral norm $$\|A\|_2 := \max_{x \neq 0} \sqrt{\frac{x^t A^t A x}{x^t x}}\,,$$

row sum norm $$\|A\|_\infty := \max_i \sum_{k=1}^{m} |a_{i,k}|\,.$$

The verification of the properties I. to IV., possible by straightforward calculations, is omitted here.

In case of an $n \times 1$-matrix a, condition IV. does not apply. Then the norm is called a **column norm**. The column norms corresponding to the examples above are the

sum norm $$\|a\|_1 = \sum_{i=1}^{n} |a_i|\,,$$

Euclidean norm $$\|a\|_2 = \sqrt{\sum_{i=1}^{n} a_i^2}\,,$$

maximum norm $$\|a\|_\infty = \max_i |a_i|\,.$$

A column and a matrix norm are said to be **consistent** if the inequality

$$\|Aa\| \leq \|A\|\ \|a\|$$

is satisfied for all A and all $\boldsymbol{a}$. The pairs of norms above with like indices are consistent. The **maximal stretching factor**

$$\sigma := \max_{\|\boldsymbol{x}\|=1} \|A\boldsymbol{x}\|$$

is a norm of A for each column norm. Examples are the norms with the indices 2 and ∞.

A sequence of $n \times m$-matrices A_k is said to be **convergent** to the limit A if the sequence of reals

$$\|A_k - A\|$$

converges to zero. Whether or not a matrix sequence converges does not depend on the choice of the norm. Also, the matrices A_k can be columns $\boldsymbol{a}_k$ or scalars a_k.

2.6 Notes and Exercises

1. The inverse of a lower triangular matrix is, if it exists, lower triangular too.
2. The diagonal matrices

$$\begin{bmatrix} \pm 1 & & & \\ & \pm 1 & & \\ & & \ddots & \\ & & & \pm 1 \end{bmatrix}$$

   are the only orthonormal triangular matrices.
3. Any two n-columns $\boldsymbol{x}$ and $\boldsymbol{y}$ satisfy the Cauchy-Schwarz inequality

$$|\boldsymbol{x}^t \boldsymbol{y}| \le \sqrt{(\boldsymbol{x}^t \boldsymbol{x}) \cdot (\boldsymbol{y}^t \boldsymbol{y})}.$$

4. The number cond $A := \|A\|_2 \cdot \|A^{-1}\|_2$ is known as the 2 - **condition number** of the non-singular matrix A. One has cond $A = 1$, if A is orthonormal and otherwise, cond $A > 1$.
5. The relative errors of the right side $\boldsymbol{a}$ and the solution $\boldsymbol{x}$ of a linear system $A\boldsymbol{x} = \boldsymbol{a}$ satisfy

$$\frac{\|\Delta \boldsymbol{x}\|_2}{\|\boldsymbol{x}\|_2} \le \text{cond}\, A \, \frac{\|\Delta \boldsymbol{a}\|_2}{\|\boldsymbol{a}\|_2}.$$

II Linear Equations and Inequalities

Many problems in applied mathematics lead to systems of linear equations or linear inequalities. One can try to solve these systems by successively improving an approximate solution or by transforming to a system that can be solved directly.

3 Gaussian Elimination

The best known method for the solution of a linear system by a transformation is **Gaussian elimination.**

3.1 Backward Substitution

Usually the linear system $A\boldsymbol{x} = \boldsymbol{a}$ has a square matrix with a non-vanishing determinant. In this case there exists exactly one solution $\boldsymbol{x}$. In certain cases this solution can be obtained immediately.

For example, if the matrix U of the linear system

$$U\boldsymbol{x} = \boldsymbol{b}$$

for the n-column $\boldsymbol{x}$ is upper-triangular

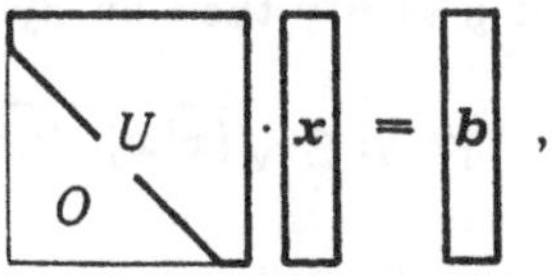

and if $u_{1,1} \cdot u_{2,2} \cdot \ldots \cdot u_{n,n} \neq 0$, one can determine x_n from the last equation, then with x_n also x_{n-1} from the second but last equation and so on. This procedure is called **backward substitution** and is described by the algorithm below:

Backward Substitution

Input: U, non-singular $n \times n$-matrix, $\boldsymbol{b} \in \mathbb{R}^n$
Output: $\boldsymbol{x} \in \mathbb{R}^n$ for which $U\boldsymbol{x} = \boldsymbol{b}$

For $k = n, n-1, \ldots, 1$
$$x_k := \frac{1}{u_{k,k}}(b_k - u_{k,k+1}x_{k+1} - \cdots - u_{k,n}x_n).$$

Falk's scheme is used to perform this algorithm by hand. Rather than computing the product Ux, the column x is calculated "backwards" from the known result $Ux = b$:

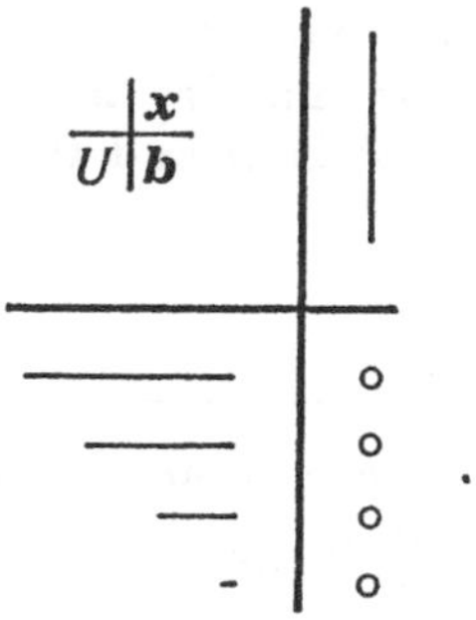

Example 1: With U and b as entered in the scheme

			x
			1
			2
			3
2	2	0	6
	−1	1	1
		2	6

one finds successively $x_3 = 3$, $x_2 = 2$ and $x_1 = 1$.

Remark 1: A linear system

$$Ly = c$$

with a non-singular lower triangular matrix L is solved analogously by **forward substitution.**

3.2 Gaussian Elimination

The solution of a linear system $Ax = a$ does not change if, among others

1. a multiple of any equation is added to another one, or
2. two equations are interchanged.

The idea to use these transformations to convert a linear system $Ax = a$ into an **equivalent** system $Ux = b$ which has the same solution and is solvable by backward substitution goes back to Gauss. It can be systematized quite simply: Using rule 1, appropriate multiples of the first row are added to all other rows in order to eliminate the coefficients of x_1 in these rows, then appropriate multiples of the new second row are added to the subsequent rows in order to eliminate the coefficients of x_2 in those rows, and so forth.

It is expedient for the execution to write down solely the matrix A and the column a rather than the entire linear system. The transformations are described by the following algorithm where the matrix A and the column a are combined in the single augmented matrix $[A, a]$:

Gauss

Input: $[A, a]$ an $n \times (n+1)$-matrix, A non-singular

Output: $[A, a] := [U, b]$, U upper triangular

1 For $j = 1, 2, \ldots, n-1$

2   if $a_{j,j} \neq 0$

3   for $i = j+1, j+2, \ldots, n$

4   subtract $\dfrac{a_{i,j}}{a_{j,j}}$ times row j from row i in $[A, a]$.

For the sake of simplicity, the intermediate coefficients are called again $a_{i,k}$. They are written over the matrix $[A, a]$ which changes to $[U, b]$ eventually.

3.3 Pivoting

The algorithm **Gauss** fails if the proposed **pivot** $a_{j,j}$ for the jth step is zero. In this case a new pivot is chosen:

Before executing step j, the jth row is interchanged with a row $r \geq j$ such that the new pivot is of the maximum modulus possible. A non-zero pivot can always be found because the rows of A remain linearly independent under Gaussian elimination. The instructions of the corresponding algorithm are the following.

Pivoting

Find $a_{r,j}$ with $|a_{r,j}| = \max_{i \geq j} |a_{i,j}|$,

transpose the rows r and j.

Again, what has been row r is referred to as row j, after the exchange. To ensure that the algorithm **Gauss** is numerically stable, a pivot is searched for even if $a_{j,j} \neq 0$. Hence, row 2 of the algorithm **Gauss** has to be replaced by

2 execute **Pivoting**

Example 2: The algorithm **Gauss** with **Pivoting** transforms the matrix $[A, \boldsymbol{a}]$ of the linear system

$$\begin{bmatrix} 2 & 2 & 0 \\ 1 & 1 & 2 \\ 2 & 1 & 1 \end{bmatrix} \begin{bmatrix} x_1 \\ x_2 \\ x_3 \end{bmatrix} = \begin{bmatrix} 6 \\ 9 \\ 7 \end{bmatrix}$$

successively into

$$\left[\begin{array}{ccc|c} 2 & 2 & 0 & 6 \\ 1 & 1 & 2 & 9 \\ 2 & 1 & 1 & 7 \end{array}\right] \overset{j=1}{\longrightarrow} \left[\begin{array}{ccc|c} 2 & 2 & 0 & 6 \\ 0 & 0 & 2 & 6 \\ 0 & -1 & 1 & 1 \end{array}\right] \overset{r=3}{\longrightarrow} \left[\begin{array}{ccc|c} 2 & 2 & 0 & 6 \\ 0 & -1 & 1 & 1 \\ 0 & 0 & 2 & 6 \end{array}\right].$$

The last step $j = 2$ is not executed since $a_{3,2}$ is already zero. The linear system

$$\begin{bmatrix} 2 & 2 & 0 \\ 0 & -1 & 1 \\ 0 & 0 & 2 \end{bmatrix} \begin{bmatrix} x_1 \\ x_2 \\ x_3 \end{bmatrix} = \begin{bmatrix} 6 \\ 1 \\ 6 \end{bmatrix}$$

has been solved by backward substitution in Example 1.

3.4 Notes and Exercises

1. Taking a pivot $a_{r,s}$ with

$$|a_{r,s}| = \max_{i,k \geq j} |a_{i,k}|$$

is called **total pivoting**. In addition to interchanging the rows r and j, the columns s and j, have to be interchanged then, too, as well as x_s and x_j. **Total pivoting** yields the most numerically stable form of the algorithm **Gauss**.

2. The algorithm **Gauss** can be accelerated by restricting the instruction **4** to the columns $k > j$ and adding the instruction

5 set $a_{i,j} := 0$.

4 The LU Factorization

The single operations of the Gauss-algorithm can be organized in different ways, one of which is a concentrated Gauss-algorithm which can be illustrated by Falk's scheme.

4.1 The LU Factorization of A

The matrix $[A, a]$ can easily be restored from the matrix $[U, b]$: Transposing the same rows in I as in A generates a matrix P, i.e., PA contains the rows of A in the transposed order. From the construction of U it is apparent then, that the ith row of U equals the ith row of PA minus a linear combination of the first $(i-1)$ rows of PA. This fact entails that, conversely, the ith row of PA equals the ith row of U plus some linear combination of the first $(i-1)$ rows of U. This property can be expressed by a lower-triangular matrix L with $l_{i,i} = 1$ and

$$LU = PA, \qquad Lb = Pa.$$

This decomposition of PA is called **LU factorization** of PA. As with forward and backward substitution, Falk's scheme facilitates determining L and U. For the sake of simplicity, $P = I$ or $LU = A$ is assumed for the moment

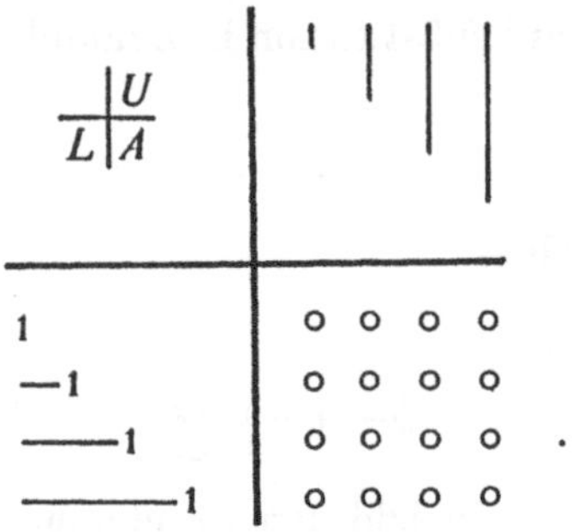

The unknowns $l_{i,j}$ and $u_{j,k}$ can be determined in some appropriate order from the scalar product

$$(\text{row } i \text{ of } L) \times (\text{column } k \text{ of } U) = a_{i,k}.$$

One can proceed column by column (Banachiewicz), row by row (Gauss) or alternating row after column (Crout). When proceeding through the columns, one determines first the first column of U, then the first column of L (without $l_{1,1} = 1$), then the second column of U and the second one of L etc., as indicated in the Figure.

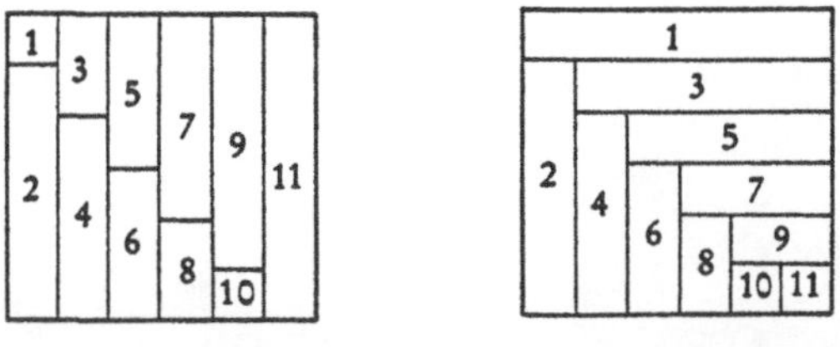

Figure 4.1 Partitioning by Banachiewicz and Crout

The method of Banachiewicz leads to the algorithm:

LU Factorization

Input: A, non-singular $n \times n$-matrix
Output: L, U with $A = LU$, $l_{i,i} = 1$

For $k = 1, 2, \ldots, n$
  for $i = 1, 2, \ldots, k$
    determine $u_{i,k} := a_{i,k} - \sum_{j=1}^{i-1} l_{i,j} u_{j,k}$,
  if $u_{k,k} \neq 0$
    for $i = k+1, k+2, \ldots, n$
      determine $l_{i,k} := \frac{1}{u_{k,k}} \left(a_{i,k} - \sum_{j=1}^{k-1} l_{i,j} u_{j,k} \right)$.
Set $u_{i,k} := l_{k,i} := 0$, $i > k$; $l_{i,i} := 1$.

One should pay heed especially to the loop of the summation index j. The entries $u_{i,k}$ of U and $l_{i,k}$ of L could be written over the $a_{i,k}$ of A. In this case the algorithm would eventually replace A by the matrix $U + L - I$.

4.2 LU Factorization with Pivoting

The algorithm **LU factorization** fails if some $u_{k,k}$ vanishes. A has no LU decomposition then, but PA does. Appropriate rows have to be interchanged. But the search for a pivot is somewhat cumbersome. The potential pivots $u_{i,k}$, $i = k, k+1, \ldots, n$, must be calculated first. However, computing the numbers $l_{i,k}$ becomes faster after the row exchange, because simply $u_{i,k}$ has to be divided by $u_{k,k}$. The variant on the LU factorization thus obtained is:

LU Factorization with Pivoting

Input: A, non-singular $n \times n$-matrix
Output: L, U with $LU = PA$ (P permutation matrix)

For $k = 1, 2, \ldots, n$
  for $i = 1, 2, \ldots, k-1$
    determine $u_{i,k} := a_{i,k} - \sum_{j=1}^{i-1} l_{i,j} u_{j,k}$.

LU Pivoting

4 For $i = k, k+1, \dots, n$

5 determine $u_{i,k} := a_{i,k} - \sum_{j=1}^{k-1} l_{i,j} u_{j,k}$,

6 find $u_{s,k}$ with $|u_{s,k}| = \max_{i \geq k} |u_{i,k}|$,

7 transpose row s and row k in L, U and A[1)]

8 For $i = k+1, k+2, \dots, n$

9 determine $l_{i,k} := \dfrac{u_{i,k}}{u_{k,k}}$.

10 Set $u_{i,k} := l_{k,i} := 0$, $i > k$, and $l_{i,i} := 1$.

Again, A can be overwritten by U and $L - I$.

Remark 1: In order to ensure that this procedure is numerically stable, a pivot search as in the algorithm **Gauss** ought to be done even in case $u_{k,k} \neq 0$. Then the size of the numbers $l_{i,k}$ calculated in row 9 is limited.

4.3 Systems of Linear Equations

Along with the LU decomposition $LU = PA$ of A goes the transformation of $A\boldsymbol{x} = \boldsymbol{a}$ into the equivalent system $U\boldsymbol{x} = \boldsymbol{b}$, where $L\boldsymbol{b} = P\boldsymbol{a}$. $P\boldsymbol{a}$ is computed on the side by extending the row interchanges in A to the augmented matrix $[A, \boldsymbol{a}]$. With this, $P\boldsymbol{a}$ will eventually take the place of $\boldsymbol{a}$. Then, $L\boldsymbol{b} = \boldsymbol{a}$ is solved through forward substitution:

Forward Substitution

11 For $i = 1, 2, \dots, n$

12 determine $b_i := a_i - \sum_{j=1}^{i-1} l_{i,j} b_j$.

Finally $U\boldsymbol{x} = \boldsymbol{b}$ is solved by the algorithm **Backward Substitution.**

Example 3: The schemes below show the LU factorization with forward substitution for the linear system of Example 2 before and after the step $k = 3$.

[1)] If A is augmented by $\boldsymbol{a}$, then also in $\boldsymbol{a}$.

$$
\begin{array}{c|c}
\dfrac{|U\,|b}{L\,|\,A\,|a} & \begin{array}{ccc|c}
2 & 2 & * & * \\
 & 0 & * & * \\
 & & * & * \\
\end{array} \\
\hline
\begin{array}{cccc}
1 & & & \\
0.5 & 1 & & \\
1 & * & 1 & \\
\end{array} &
\begin{array}{ccc|c}
2 & 2 & 0 & 6 \\
1 & 1 & 2 & 9 \\
2 & 1 & 1 & 7 \\
\end{array}
\end{array}
\quad \xrightarrow{\ k=3\ } \quad
\begin{array}{c|c}
& \begin{array}{ccc|c}
2 & 2 & 0 & 6 \\
 & -1 & 1 & 1 \\
 & & 2 & 6 \\
\end{array} \\
\hline
\begin{array}{cccc}
1 & & & \\
1 & 1 & & \\
0.5 & 0 & 1 & \\
\end{array} &
\begin{array}{ccc|c}
2 & 2 & 0 & 6 \\
2 & 1 & 1 & 7 \\
1 & 1 & 2 & 9 \\
\end{array}
\end{array}
$$

The values which are not determined before step $k = 3$ are marked by *.

4.4 Notes and Exercises

1. A square matrix A has exactly one LU factorization (without row interchanges) if all leading principal submatrices

$$
\begin{bmatrix}
a_{1,1} & \cdots & a_{1,k} \\
\vdots & & \vdots \\
a_{k,1} & \cdots & a_{k,k}
\end{bmatrix}, \qquad k = 1,\ldots,n,
$$

   are non-singular.

2. The LU factorization (without row interchanges) of A is unique in case it exists.

3. The numbers $u_{i,k}$ and $l_{i,k}$ (except the elements $l_{i,i} = 1$) which are computed during the LU factorization of A with or without pivoting can be written over the entries $a_{i,k}$ of A. This minimizes the use of space in analogy to the algorithm **Gauss**.

5 The Exchange Algorithm

There is a third form of Gauss-algorithm of interest for the direct solution of a linear system which is also used in Chapter **10** to solve linear programs.

5.1 Exchanging Variables

Rather than considering a system of linear equations one thinks of $y = Ax$ as a linear relation with an $n \times m$-matrix A. Explicitly:

$$
\begin{aligned}
y_1 &= a_{1,1}x_1 + \cdots + a_{1,3}x_3 + \cdots + a_{1,m}x_m \\
y_2 &= a_{2,1}x_1 + \cdots + a_{2,3}x_3 + \cdots + a_{2,m}x_m \\
&\ \vdots \qquad \vdots \qquad\qquad \vdots \qquad\qquad \vdots \\
y_n &= a_{n,1}x_1 + \cdots + a_{n,3}x_3 + \cdots + a_{n,m}x_m .
\end{aligned}
$$

In case that, for instance, $a_{2,3}$ is non-zero, the equation for y_2 can be solved for x_3:

$$x_3 = -\frac{a_{2,1}}{a_{2,3}}x_1 - \cdots + \frac{1}{a_{2,3}}y_2 - \cdots - \frac{a_{2,m}}{a_{2,3}}x_m, \tag{1}$$

and the expression for x_3 so obtained can be substituted in all the other equations, e.g., the first equation becomes

$$\begin{aligned} y_1 = \quad & a_{1,1}x_1 + \cdots + 0 \cdot x_3 + \cdots + \quad a_{1,m}x_m \\ & - \frac{a_{1,3}}{a_{2,3}}a_{2,1}x_1 - \cdots + \frac{a_{1,3}}{a_{2,3}}y_2 - \cdots - \frac{a_{1,3}}{a_{2,3}}a_{2,m}x_m \,. \end{aligned}$$

Analogous equations hold for the other y_i except for the equation of y_2 which is to be replaced by (1). Thus the variables y_2 and x_3 are exchanged and the linear relation $y = Ax$ can be written as

$$y' = A'x' ,$$

where

$$y' = \begin{bmatrix} y_1 \\ x_3 \\ y_3 \\ \vdots \\ y_n \end{bmatrix}, \qquad x' = \begin{bmatrix} x_1 \\ x_2 \\ y_2 \\ x_4 \\ \vdots \\ x_m \end{bmatrix}.$$

A' is the matrix of the new coefficients.

5.2 Scheme and Algorithm

The computation of the new coefficients is carried out best in the rectangular scheme of the numbers $a_{i,k}$; the y_i are written left of it, the x_k above:

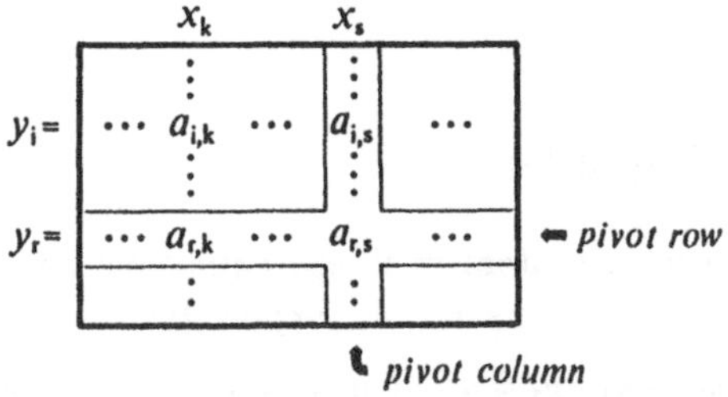

In case $a_{r,s}$ is non-zero, y_r and x_s can be exchanged. Then, the entry $a_{r,s}$ is called the **pivot**, the column s **pivot column** and the row r **pivot row**. The exchange is done as in the example above. In general, the algorithm for the exchange of y_r and x_s takes the form:

Exchange r, s

Input: A, $n \times m$-matrix where $a_{r,s} \neq 0$
Output: A after substituting the variable x_s by y_r

```
    general:
1   For i = 1, 2, ..., n, but i ≠ r,
2       for k = 1, 2, ..., m, but k ≠ s,
3           determine a_{i,k} := a_{i,k} - a_{i,s} a_{r,k} / a_{r,s} .
    pivot row:
4   For k = 1, 2, ..., m, but k ≠ s,
5       determine a_{r,k} := - a_{r,k} / a_{r,s} .
    pivot column:
6   For i = 1, 2, ..., n, but i ≠ r,
7       determine a_{i,s} := a_{i,s} / a_{r,s} .
    pivot:
8   Set a_{r,s} := 1 / a_{r,s} .
```

A is overwritten by the new matrix. The index pair r, s should be saved.

Remark 1: Instruction **3** is memorized more easily in the form

$$a_{i,k} := \frac{1}{a_{r,s}} \begin{vmatrix} a_{i,k} & a_{i,s} \\ a_{r,k} & a_{r,s} \end{vmatrix} .$$

The coefficients involved form a rectangle obtained from the intersection of the rows i, r with the columns k, s.

Example 1: For $r = 1$, $s = 1$

$$\begin{array}{c} \\ y_1 = \\ y_2 = \\ y_3 = \end{array} \begin{array}{|ccc|} x_1 & x_2 & x_3 \\ \hline \underline{2} & 2 & 0 \\ 1 & 1 & 2 \\ 2 & 1 & 1 \\ \hline \end{array} \quad \text{is transformed into} \quad \begin{array}{c} \\ x_1 = \\ y_2 = \\ y_3 = \end{array} \begin{array}{|ccc|} y_1 & x_2 & x_3 \\ \hline 0.5 & -1 & 0 \\ 0.5 & 0 & 2 \\ 1 & -1 & 1 \\ \hline \end{array} .$$

The pivot is marked.

5.3 Inversion

In case of a non-singular matrix A one can interchange all y_i with all x_k. This leads after n steps to the transformation inverse to $y = Ax$. If the pivots can be taken from the diagonal, one obtains A^{-1}, the inverse of A.

However, a maximal pivot should be chosen, anyway, to ensure numerical stability. In general, the order of both the y_i and x_k is destroyed by **total** or even **partial pivoting**. One can circumvent the associated permutations of the matrix by installing an index transformation. For better readability, though, the algorithm is presented without such an index transformation:

Inversion with Total Pivoting

Input:	A, non-singular $n \times n$-matrix
Output:	$A := A^{-1}$ inverse to A

1 For $j = 1, 2, \ldots, n$

2 execute **Total Pivoting**

3 Find $r, s \geq j$ with $|a_{r,s}| = \max\limits_{i,k \geq j} |a_{i,k}|$,

4 interchange the rows j and r
and the columns j and s of A .

5 execute **Exchange j, j** .

6 Restore the natural order of the rows and columns.

Example 2: The matrix of Example 1 is successively modified by the algorithm **Inversion** as follows:

$$\begin{array}{c|ccc|} & y_1 & x_3 & x_2 \\ \hline x_1 = & 0.5 & 0 & -1 \\ y_2 = & 0.5 & \underline{2} & 0 \\ y_3 = & 1 & 1 & -1 \\ \hline \end{array}\ , \quad \begin{array}{c|ccc|} & y_1 & y_2 & x_2 \\ \hline x_1 = & 0.5 & 0 & -1 \\ x_3 = & -0.25 & 0.5 & 0 \\ y_3 = & 0.75 & 0.5 & \underline{-1} \\ \hline \end{array}\ , \quad \begin{array}{c|ccc|} & y_1 & y_2 & y_3 \\ \hline x_1 = & -0.25 & -0.5 & 1 \\ x_3 = & -0.25 & 0.5 & 0 \\ x_2 = & 0.75 & 0.5 & -1 \\ \hline \end{array}\ .$$

The pivots are marked. After reordering, the matrix

$$\begin{bmatrix} 2 & 2 & 0 \\ 1 & 1 & 2 \\ 2 & 1 & 1 \end{bmatrix}^{-1} = \begin{bmatrix} -0.25 & -0.5 & 1 \\ 0.75 & 0.5 & -1 \\ -0.25 & 0.5 & 0 \end{bmatrix}$$

is obtained.

5.4 Linear Equations

The exchange algorithm can also be used to solve the linear system $A\boldsymbol{x} + \boldsymbol{a} = \mathbf{o}$[1]. One sets $\boldsymbol{y} = A\boldsymbol{x} + \boldsymbol{a} \cdot 1 = \mathbf{o}$, i.e., the scheme 5.1 is simply augmented by the column $\boldsymbol{a}$:

$$
\begin{array}{r|ccccc|c|}
 & & x_k & & x_s & & 1 \\
\hline
 & & \vdots & & \vdots & & \vdots \\
y_i = & \cdots & a_{i,k} & \cdots & a_{i,s} & \cdots & a_i \\
 & & \vdots & & \vdots & & \vdots \\
y_r = & \cdots & a_{r,k} & \cdots & a_{r,s} & \cdots & a_r \\
 & & \vdots & & \vdots & & \vdots \\
\hline
\end{array} \; .
$$

Then, $\boldsymbol{x}$ is the solution of the linear system for $\boldsymbol{y} = \mathbf{o}$. The column headed by 1, therefore, presents the solution after exchanging all y_i with x_k. The pivot columns need not to be determined because the y_i are set to zero.

Here, too, one cannot do without pivoting but it is not necessary to order the rows in the end, if the order of the x_k is not destroyed. Partial instead of total pivoting keeps the algorithm simple. This form of the algorithm is the following:

Gauss with Exchange

Input: $[A, \boldsymbol{a}]$, A non-singular $n \times n$-matrix, $\boldsymbol{a}$ n-column
Output: $[A, \boldsymbol{a}] := [*, \boldsymbol{x}]$ with $A\boldsymbol{x} + \boldsymbol{a} = \mathbf{o}$

```
For j = 1, 2, ..., n
    execute Partial Pivoting
        Determine a_{r,j} with |a_{r,j}| = max_{i≥j} |a_{i,j}| ,
        interchange the rows j and r of [A, a] .
    for k = j+1, j+2, ..., n+1
        for i = 1, 2, ..., n, but i ≠ j,
            set a_{i,k} := a_{i,k} - a_{i,j} a_{j,k} / a_{j,j} ,
        set a_{j,k} := - a_{j,k} / a_{j,j} .
```

The entries of the column $\boldsymbol{a}$ are denoted by $a_{i,n+1}$ for the sake of simplicity. The values $*$ are not calculated.

---

[1] Observe that $A\boldsymbol{x} = \boldsymbol{a}$ is not written as usual.

Example 3: **The performance of the algorithm Gauss with Exchange is illustrated for the Example 2 in 3.3:**

$$
\begin{array}{r|ccc|c|}
 & x_1 & x_2 & x_3 & 1 \\ \hline
y_1 = & \underline{2} & 2 & 0 & -6 \\
y_2 = & 1 & 1 & 2 & -9 \\
y_3 = & 2 & 1 & 1 & -7 \\ \hline
\end{array}
\quad \xrightarrow{j=1} \quad
\begin{array}{r|ccc|c|}
 & 0 & x_2 & x_3 & 1 \\ \hline
x_1 = & * & -1 & 0 & 3 \\
y_2 = & * & 0 & 2 & -6 \\
y_3 = & * & \underline{-1} & 1 & -1 \\ \hline
\end{array}
\quad \Big\rangle\, r=3
$$

$$
\begin{array}{r|ccc|c|}
 & 0 & x_2 & x_3 & 1 \\ \hline
x_1 = & * & -1 & 0 & 3 \\
y_3 = & * & \underline{-1} & 1 & -1 \\
y_2 = & * & 0 & 2 & -6 \\ \hline
\end{array}
\quad \xrightarrow{j=2} \quad
\begin{array}{r|ccc|c|}
 & 0 & 0 & x_3 & 1 \\ \hline
x_1 = & * & * & -1 & 4 \\
x_2 = & * & * & 1 & -1 \\
y_2 = & * & * & \underline{2} & -6 \\ \hline
\end{array}
\quad \xrightarrow{j=3}
$$

$$
\begin{array}{r|ccc|c|}
 & 0 & 0 & 0 & 1 \\ \hline
x_1 = & * & * & * & 1 \\
x_2 = & * & * & * & 2 \\
x_3 = & * & * & * & 3 \\ \hline
\end{array}
\quad . \qquad \text{Thus } x = \begin{bmatrix} 1 \\ 2 \\ 3 \end{bmatrix} .
$$

The pivots are underlined.

5.5 Notes and Exercises

1. **A non-vanishing column pivot always exists if A is non-singular.**
2. **The submatrices**

$$
\left[\begin{array}{ccc|c}
a_{j,j} & \cdots & a_{j,n} & a_j \\
\vdots & & \vdots & \vdots \\
a_{n,j} & \cdots & a_{n,n} & a_n
\end{array}\right]
$$

of the matrix $[A, a]$ as updated in the jth steps of the algorithm Gauss and Gauss with Exchange coincide in the examples and in general, why?

6 The Cholesky Factorization

In many applications the matrix of a linear system $A\boldsymbol{x} = \boldsymbol{a}$ is symmetric. But Gaussian eliminations, the LU factorization and the exchange algorithm destroy the symmetry. However, these methods can be modified so that the symmetry is maintained and even advantage taken of.

6.1 Symmetrical Factorization

A symmetric matrix A with $\boldsymbol{x}^t A\boldsymbol{x} > 0$ for all $\boldsymbol{x} \neq \mathbf{o}$ is called **positive definite.** For the LU factorization of a symmetric, positive definite matrix A one looks for an upper-triangular matrix C which satisfies

$$A = C^t C$$

and tries its computation in Falk's scheme:

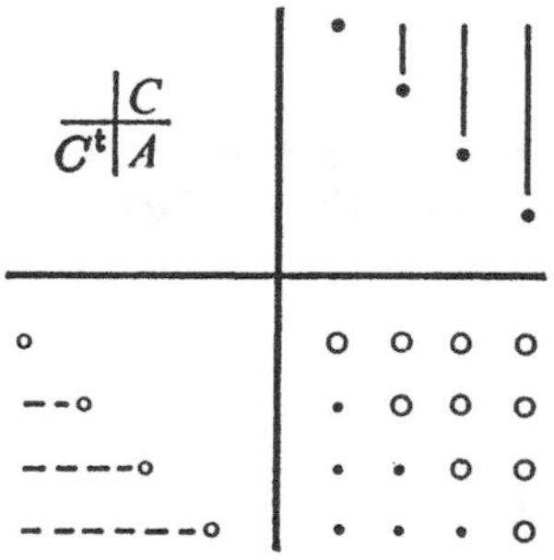

The entries of A below the diagonal and of C^t need not be listed because of the symmetry. The $c_{i,k}$ for $i \leq k$ are successively determined from the equation

$$(\text{column } i \text{ of } C) \times (\text{column } k \text{ of } C) = a_{i,k}\,.$$

The term $c_{k,k}^2$ occurs for $i = k$ and the square root must be taken. In **6.2** it is shown that $c_{k,k}^2 > 0$ if A is positive definite.

Proceeding through A column by column down to the diagonal establishes the algorithm:

[1] The instructions **2** and **3** are void for $k = 1$.

Cholesky

Input:	A, symmetric positive definite $n \times n$-matrix
Output:	C upper triangular matrix with $C^tC = A$

For $k = 1, 2, \dots, n$

  for $i = 1, 2, \dots, k-1$[1)]

    determine $c_{i,k} := \frac{1}{c_{i,i}}\left(a_{i,k} - \sum_{j=1}^{i-1} c_{j,i}c_{j,k}\right)$.

  Determine $c_{k,k} := {}_{+}\sqrt{a_{k,k} - \sum_{j=1}^{k-1} c_{j,k}^2}$.

  For $i = k+1, k+2, \dots, n$

    set $c_{i,k} := 0$.

Pivoting is not necessary. The Cholesky factorization of positive definite matrices is numerically stable.

Remark 1: Making use of the symmetry reduces the amount of computation. It is approximately halved compared with the LU factorization; n square roots must be taken additionally, though.

6.2 Existence and Uniqueness

The algorithm **Cholesky** would fail if some $c_{k,k}^2 \leq 0$. Fortunately, there is

Theorem 1: *If $A = A^T$ is positive definite, then all $c_{k,k}^2$ are positive. The Cholesky decomposition is unique if positive square roots are taken.*

For the proof assume that the algorithm failed for the first time because of $c_{j,j}^2 \leq 0$. Then a $y \neq o$ can be found with $y^tAy = c_{j,j}^2 \leq 0$, namely the solution y of the linear system

$$Ry = b\,,$$

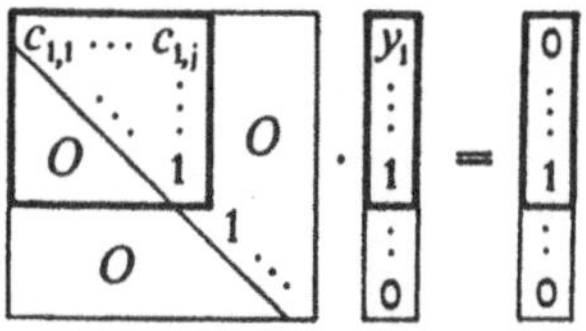

,

where $r_{i,k} := c_{i,k}$ for $i<j$ and $i \le k \le j$, $r_{k,k} := 1$ for $k \ge j$, $b_j := 1$ and $r_{i,k} := b_i := 0$ otherwise. Then $y_j = 1$ ensues and

$$y^t A y = y^t R^t R y - 1 + c_{j,j}^2 = b^t b - 1 + c_{j,j}^2 = c_{j,j}^2 \le 0 .$$

This contradicts the stipulation that A is positive definite.

If there is a second factorization $B^t B = A = C^t C$ with positive diagonal elements $b_{k,k}$, then

$$I = [B^t]^{-1} C^t C B^{-1} = [CB^{-1}]^t [CB^{-1}]$$

would be another different Cholesky factorization of the identity matrix besides $I = I \cdot I$. Since the factorization of I is unique, B must equal C.

6.3 Symmetric Systems of Linear Equations

The symmetric system of linear equations $A\boldsymbol{x} = \boldsymbol{a}$ with a positive definite matrix A and the corresponding Cholesky factorization $A = C^t C$ is equivalent to the interlocking system

$$C\boldsymbol{x} = \boldsymbol{b} \quad \text{where} \quad C^t \boldsymbol{b} = \boldsymbol{a} .$$

First, $\boldsymbol{b}$ is determined through forward substitution (one should observe that in general $c_{i,i} \ne 1$). Then, $C\boldsymbol{x} = \boldsymbol{b}$ is solved by backward substitution where $U = C$.

Example 1: The rudimentary scheme of Falk in **6.1** for the symmetric linear system

$$\begin{bmatrix} 1 & 1 & -2 \\ 1 & 5 & 0 \\ -2 & 0 & 14 \end{bmatrix} \begin{bmatrix} x_1 \\ x_2 \\ x_3 \end{bmatrix} = \begin{bmatrix} 3 \\ 13 \\ 8 \end{bmatrix}$$

has the form

$$\begin{array}{c|c} C & b \\ \hline A & a \end{array} \qquad \begin{array}{ccc|c} 1 & 1 & -2 & 3 \\ & 2 & 1 & 5 \\ & & 3 & 3 \\ \hline 1 & 1 & -2 & 3 \\ \cdot & 5 & 0 & 13 \\ \cdot & \cdot & 14 & 8 \end{array} .$$

This means that the algorithm **Cholesky** leads to the equivalent system

$$\begin{bmatrix} 1 & 1 & -2 \\ 0 & 2 & 1 \\ 0 & 0 & 3 \end{bmatrix} \begin{bmatrix} x_1 \\ x_2 \\ x_3 \end{bmatrix} = \begin{bmatrix} 3 \\ 5 \\ 3 \end{bmatrix} \quad \text{with the solution} \quad \boldsymbol{x} = \begin{bmatrix} 3 \\ 2 \\ 1 \end{bmatrix} .$$

6.4 Iterative Refinement

The solution $\boldsymbol{p}$ of the linear system $A\boldsymbol{x} = \boldsymbol{a}$ obtained by means of the LU factorization (or the Cholesky factorization) may not be sufficiently close to the exact solution, i.e., the **residual**

$$\boldsymbol{r} := A\boldsymbol{p} - \boldsymbol{a}$$

differs clearly from $\mathbf{o}$. In general such a solution can be improved subsequently by an **iterative refinement**. Let $\boldsymbol{d}$ be the **deviation** from the exact solution, i.e., $\boldsymbol{x} = \boldsymbol{p} - \boldsymbol{d}$. From the linear system one can conclude the fact

$$A\boldsymbol{d} = \boldsymbol{r}\,.$$

If the decomposition $A = LU$ is available, then $\boldsymbol{d}$ can be determined by forward and backward substitution from the linear systems

$$L\boldsymbol{q} = \boldsymbol{r} \quad \text{and} \quad U\boldsymbol{d} = \boldsymbol{q}\,.$$

For this, the Falk scheme is augmented by the columns $\boldsymbol{r}$ and $\boldsymbol{q}$. The residual $\boldsymbol{r}$ can be used effectively to correct $\boldsymbol{p}$ only if it is computed more precisely than $\boldsymbol{p}$. Therefore, one determines $\boldsymbol{r} = A\boldsymbol{p} - \boldsymbol{a}$ with **double precision.**

Example 2: The linear system

$$\begin{bmatrix} 2 & 1 & 0 \\ 1 & 4 & 1 \\ 0 & 1 & 2 \end{bmatrix} \begin{bmatrix} x_1 \\ x_2 \\ x_3 \end{bmatrix} = \begin{bmatrix} 2 \\ 8 \\ 2 \end{bmatrix}$$

is solved computing with two significant digits. The respective extended rudimentary scheme is:

	$\boldsymbol{p}$	$\boldsymbol{d}$
C	$\boldsymbol{b}$	$\boldsymbol{q}$
A	$\boldsymbol{a}$	$\boldsymbol{r}$

			0.0	0.01
			2.1	0.083
			−0.11	−0.053
1.4	0.71	0.0	1.4	0.071
	1.8	0.55	3.8	0.13
		1.3	−0.15	−0.07
2	1	0	2	0.1
·	4	1	8	0.29
·	·	2	2	−0.02

The exact solution $\boldsymbol{x} = \begin{bmatrix} 0 \\ 2 \\ 0 \end{bmatrix}$ is almost reached by $\boldsymbol{p} - \boldsymbol{d} = \begin{bmatrix} -0.01 \\ 2.0 \\ -0.06 \end{bmatrix}$.

6.5 Notes and Exercises

1. The linear mapping $\boldsymbol{y} = C\boldsymbol{x}$ transforms the quadratic form $\boldsymbol{x}^t A \boldsymbol{x}$ with $A = C^t C$ into $\boldsymbol{y}^t \boldsymbol{y}$ (Cholesky 1871).
2. The matrix C as generated by the algorithm Cholesky can also be generated by the algorithm Gauss if the jth row is divided by the square root of the diagonal element before each step j, $j = 1, 2, \ldots, n$.
3. The diagonal of a positive definite matrix is positive.
4. The largest entry of a positive definite matrix lies on the main diagonal.
5. Every symmetric matrix A with positive entries on the diagonal satisfying the **strong column sum criterion**

$$2\,|a_{k,k}| > \sum_{i=1}^{n} |a_{i,k}| \quad \text{for all } k$$

is positive definite.

7 The QU Factorization

The various forms of the Gauss-algorithm can change the condition of a linear system. This can be avoided if the matrix A is transformed into a triangular matrix U by multiplication with an orthonormal matrix Q^t. This then implies $A = QU$.

7.1 The Householder Transformation

The modern form of the QU decomposition of an $n \times m$-matrix A, $n \geq m$, rank $A = m$, going back to Householder uses orthogonal transformations of the form

$$\boldsymbol{y} = H\boldsymbol{x} \quad \text{with} \quad H = I - \boldsymbol{u} \cdot 2\boldsymbol{u}^t, \quad \boldsymbol{u}^t\boldsymbol{u} = 1\,.$$

These transformations are set up to map a certain column $\boldsymbol{a} = [0, \ldots, 0, a_i, \ldots, a_n]^t$ to a multiple of the ith coordinate vector $\boldsymbol{e}_i = [0, \ldots, 1, \ldots, 0]^t$. One easily verifies: H is symmetric, involutory and therefore, orthonormal, i.e.,

$$H^t = H, \quad HH = I \quad \text{and thus} \quad H^t H = I\,.$$

Determining $\boldsymbol{u}$ from $\boldsymbol{a}$ and $\boldsymbol{e}_i$ is simple: From the constraint

$$H\boldsymbol{a} = \boldsymbol{a} - \boldsymbol{u} \cdot 2\boldsymbol{u}^t\boldsymbol{a} = \boldsymbol{e}_i\gamma$$

follows $\gamma = \pm\|a\|_2$ for H is orthonormal, and, furthermore, since $2u^t a$ is a scalar, that u equals the normalized vector $v := a - e_i\gamma$, i.e.,

$$u = \frac{a - e_i\gamma}{\|a - e_i\gamma\|_2}.$$

To avoid cancellation if a is nearly parallel to e_i one sets

$$\gamma := -\text{sign}\; a_i \,\|a\|_2 \neq 0^{1)}.$$

One should observe that v differs from a only in the ith coordinate a_i and that

$$u2u^t = v\frac{1}{\alpha}v^t \quad \text{with} \quad \alpha := \frac{1}{2}v^t v = \gamma^2 - a_i\gamma.$$

holds. Thus, the reflection is given by

$$y = Hx = x - v \cdot s \quad \text{with} \quad s := \frac{1}{\alpha}v^t x. \tag{1}$$

It is useful to know that Hw and w coincide in the first $i-1$ components for any column $w = [w_1, \ldots, w_n]^t$ and that in particular $Hw = w$ if $w_i = \cdots = w_n = 0$.

7.2 The Householder Algorithm

One can transform any $n \times m$-matrix A, $n \geq m$, successively into the first m columns of an upper triangular matrix U through multiplication by m Householder transformations H_j, $j = 1, 2, \ldots, m$ (or $n-1$ if $m = n$), where H_1 maps the first column a_1 of A to e_1, H_2 the vector $\bar{a}_2$ which is the second column a_2 of $H_1 A$ after setting all components above the diagonal to zero, to e_2, and so forth.

It would be expensive to compute the matrices H_j of the Householder transformations and to multiply them. .Instead, one employs identity (1) while taking advantage of the observation following (1) to determine $A := H_j A$ by applying H_j to each column of A. It is common to write the entries $u_{i,k}$ over the $a_{i,k}$. This is done for each $j = 1, \ldots, m$ (or $n-1$). The corresponding algorithm with the formulas in **7.1** and the simplification $\bar{a}_k := [a_{j,k}, \ldots, a_{n,k}]^t$ (which depends on j) is then:

1) In case $a_i = 0$ one sets $\gamma := -\|a\|_2$.

Householder

Input:	A, $n \times m$-matrix, $n \geq m$, rank $A = m$
Output:	$A := Q^t A$ upper triangular matrix with $Q^t Q = I$

For $j = 1, 2, \ldots, m$ (or $j = 1, 2, \ldots, n-1$)
  determine $a^2 := \bar{a}_j^t \bar{a}_j$
  and $\gamma := -\text{sign}\, a_{j,j} \cdot {}_+\sqrt{a^2}$
  and $\alpha := a^2 - a_{j,j} \cdot \gamma$.
  Set $a_{j,j} := a_{j,j} - \gamma$, i.e., replace $\bar{a}_j$ by $\bar{v}$.
  For $k = j+1, j+2, \ldots, m$
    determine $s_k := \frac{1}{\alpha} \bar{a}_j^t \bar{a}_k$
    and set $\bar{a}_k := \bar{a}_k - \bar{a}_j s_k$.
  Set $\bar{a}_j^t := [\gamma, 0, \ldots, 0]$.

The calculation on paper, however, is better organized by the scheme

$$\begin{array}{c|c} \boldsymbol{v} & A \\ \hline \alpha & \boldsymbol{s}^t \end{array} \quad \longrightarrow \quad A' ,$$

where A' presents the new matrix $H_j A$ and the row $\boldsymbol{s}^t$ the s_k. The scheme of step j can be restricted to the $a_{i,k}$ with $i, k \geq j$.

Remark 1: The algorithm **Householder** determines the matrix U of the decomposition $A = QU$ where Q is an orthonormal $n \times n$-matrix:

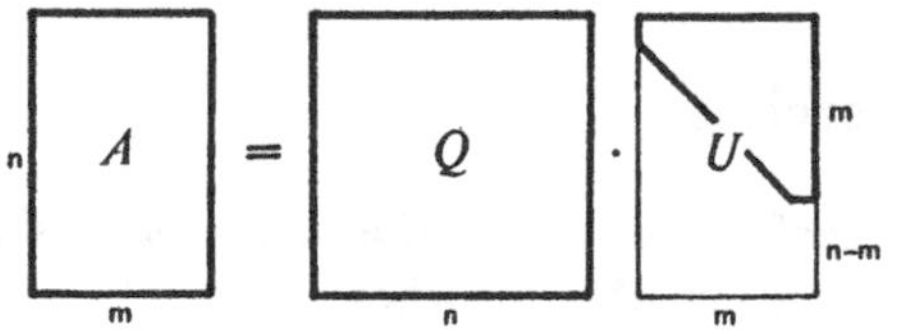

7.3 Systems of Linear Equations

The product of the Householder matrices

$$H = H_{n-1} \cdots H_2 H_1$$

transforms a linear system $Ax = a$ with an $n \times n$-matrix A into a system with an upper triangular matrix U:

$$H(Ax - a) = HAx - Ha = Ux - b.$$

Obviously, $b := Ha$ is obtained on the side as in the algorithm **Gauss**. A must just be augmented by a and the instructions 7 and 8 of the algorithm extended to $a_{n+1} := a$.

Example 1: The columns of the linear system

$$\begin{bmatrix} 1 & 1.1 & 1.1 \\ 1 & 0.9 & 0.9 \\ 0 & -0.1 & 0.2 \end{bmatrix} \begin{bmatrix} x_1 \\ x_2 \\ x_3 \end{bmatrix} = \begin{bmatrix} 1 \\ 1 \\ 0.3 \end{bmatrix}$$

are almost linearly dependent. The two steps of the two Householder transformations H_j, $j = 1, 2$, are

2,41	1	1.1	1.1	1	$\xrightarrow{j=1}$	−1.41	−1.41	−1.41	−1.41
1	1	0.9	0.9	1		0	−0.14	−0.14	0
0	0	−0.1	0.2	0.3		0	−0.1	0.2	0.3
3.41	1	1.04	1.04	1.0					

−0.31	*	−0.14	−0.14	0	$\xrightarrow{j=2}$	*	0.17	−0.02	−0.19
−0.1	*	−0.1	0.2	0.3		*	0	0.24	0.24
0.05	*	1	0.40	−0.60					

The new linear system is

$$\begin{bmatrix} -1.41 & -1.41 & -1.41 \\ 0 & 0.17 & -0.02 \\ 0 & 0 & 0.24 \end{bmatrix} \begin{bmatrix} x_1 \\ x_2 \\ x_3 \end{bmatrix} = \begin{bmatrix} -1.41 \\ -0.19 \\ 0.24 \end{bmatrix}$$

and has the solution $x = [1, -1, 1]^t$.

7.4 Notes and Exercises

1. The Householder transformation $y = Hx = x - u \cdot 2u^t x$ is a reflection about the plane with the equation $u^t x = 0$ where $u^t u = 1$.

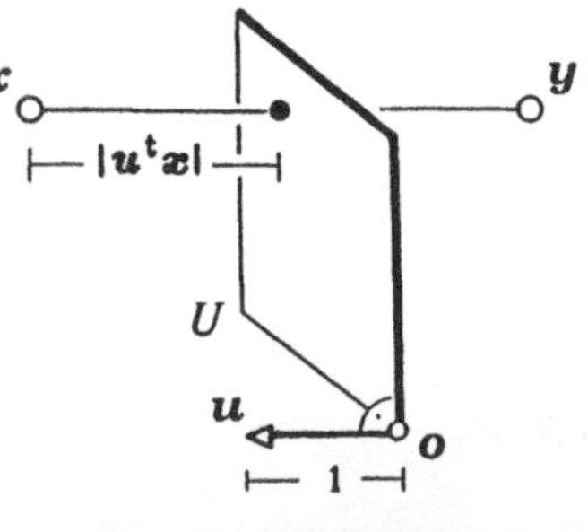

Figure 7.1
Reflection after Householder

2. The columns $\mathbf{q}_1, \ldots, \mathbf{q}_n$ of Q^t can be obtained, for instance, by taking the columns $\mathbf{e}_1, \ldots, \mathbf{e}_n$ of I for $\mathbf{a}$ and transforming them in the same way as $\mathbf{a}$.
3. The direct QU factorization of A in Falk's scheme leads to the **Gram-Schmidt orthogonalization**. It has the disadvantage to be not numerically stable without additional measures if the columns of A are nearly linearly dependent. This process differs from the Householder algorithm in that it determines Q and U of the form

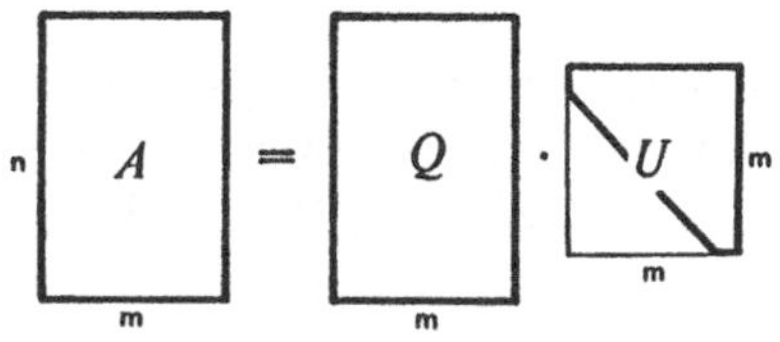

.

8 Relaxation Methods

For large sparse linear systems, there are more effective methods than the direct methods just described. These are the iterative methods which leave A unchanged and improve an approximate solution $\mathbf{p}$ iteratively.

8.1 Coordinate Relaxation

Let $\mathbf{p}$ be an approximate solution for the linear system $A\mathbf{x} = \mathbf{a}$ with the residual

$$\mathbf{r} = A\mathbf{p} - \mathbf{a} \neq 0 .$$

The basic idea of **relaxation**, namely to change $\mathbf{p}$ such that $\mathbf{r}$ becomes smaller (in some appropriate sense), goes also back to Gauss.

It is especially simple, but effective, to change merely one component p_j of $\mathbf{p}$, i.e., to replace p_j by

$$p_j' := p_j + \varrho$$

and to choose ϱ such that the corresponding component $r_j' := r_j + a_{j,j}\varrho$ of the new residual vanishes, this means

$$\varrho := -\frac{r_j}{a_{j,j}} .$$

The new solution $\mathbf{p}' = \mathbf{p} + \mathbf{e}_j \varrho$ has then the residual

$$\mathbf{r}' = \mathbf{r} + \mathbf{a}_j \varrho ;$$

$\boldsymbol{a}_j$ is the jth column of the matrix A and $\boldsymbol{e}_j$ the jth column of I.

There are several strategies for choosing the component r_j to be made zero.

Manual relaxation: Manually it is easy to select the p_j associated with the component r_j of maximal absolute value of the residual:

$$|r_j| = \max_i |r_i| ,$$

and to change the approximate solution accordingly. This is repeated until all components r_i are of lesser absolute value than a given tolerance σ_0. The corresponding algorithm works for any matrix A satisfying the conditions of Theorem 1 or 3 in **8.2** and **8.4**:

Relaxation

Input: A, non-singular $n \times n$-matrix, $a_{i,i} \neq 0$, for $i = 1, \ldots, n$;
$\boldsymbol{a}, \boldsymbol{p}$ n-columns; $\sigma_0 > 0$ tolerance

Output: $\boldsymbol{p} := \boldsymbol{p}'$ with $\|A\boldsymbol{p}' - \boldsymbol{a}\|_\infty \leq \sigma_0$

1 Set $k := 0$, $\sigma := \sigma_k$.

2 Determine $\boldsymbol{r} := A\boldsymbol{p} - \boldsymbol{a}$ (residual).

Selection of r_j for **manual relaxation**

3 Select r_j with $|r_j| = \max_i |r_i|$.

4 If $|r_j| \leq \sigma$: **stop** .

7 Set $\varrho := \dfrac{-r_j}{a_{j,j}}$,

8 $p_j := p_j + \varrho$,

9 and $\boldsymbol{r} := \boldsymbol{r} + \boldsymbol{a}_j \varrho$.

10 Go to **3**.

The algorithm is set up so as to allow for the replacement of the part "r_j for **manual relaxation**".

Cyclic relaxation: The manual relaxation is unsuitable for automation. Searching for the maximal $|r_j|$ is in most cases more costly than a step of the relaxation itself. Therefore, one lets j run simply from 1 to n. This method is known as the **Gauss-Seidel relaxation.** It uses the following algorithm for the selection of r_j:

r_j for **cyclic relaxation**

3	If $\|\|r\|\|_\infty \le \sigma$: **stop** .
4	For $j = 1, 2, \ldots, n$

It can be put directly into the algorithm **Relaxation.**

Threshold relaxation: A step of the cyclic relaxation is carried out even if

$$|r_j| \ll \max_i |r_i|$$

It sounds reasonable to change p_j in a cycle only if the corresponding $|r_j|$ is bigger than some **threshold value** σ. The relaxation terminates if all $|r_i| \le \sigma$. One can circumvent all disadvantages of the cyclic relaxation with a series of threshold values $\sigma_0 > \sigma_1 > \cdots > \sigma_m$ which take the place of σ one after another. The algorithm for the selection of r_j in this case is:

r_j for **threshold relaxation**

3	If $\|\|r\|\|_\infty \le \sigma$ and $k = m$: **stop** .
4	$k := k + 1,\ \sigma := \sigma_k$.
5	For $j = 1, 2, \ldots, n$
6	if $\|r_j\| > \sigma$

This part, too, can be put directly into the algorithm **Relaxation**. The instructions **7, 8, 9** are subordinated to instruction **6** and the tolerance σ_0 in the declaration part has to be replaced by the threshold values $\sigma_0 > \cdots > \sigma_m$.

8.2 Convergence for Diagonally Dominant Matrices

The coordinate relaxation does not necessarily force the residual to become small. Whether or not it does depends on A.

The matrix A is called **diagonally dominant** if its columns a_k satisfy the **strong column sum criterion**

$$\text{(1)} \qquad \sum_{\substack{i=1\\ i\neq k}}^{n} |a_{i,k}| < |a_{k,k}|, \quad k = 1, \ldots, n.$$

Theorem 1: *Manual, cyclic and threshold relaxation converge for all diagonally dominant matrices A.*

To begin the proof, the inequality $\|\boldsymbol{r}'\|_1 \leq \|\boldsymbol{r}\|_1$ is derived first. Because $\boldsymbol{r}' = \boldsymbol{r} + \boldsymbol{a}_j\varrho$, $\varrho = -r_j/a_{j,j}$ and $r'_j = 0$ it follows that

$$\|\boldsymbol{r}'\|_1 = \sum_{\substack{i=1\\i\neq j}}^{n} |r_i + a_{i,j}\varrho| \leq \sum_{\substack{i=1\\i\neq j}}^{n} |r_i| + |\varrho| \sum_{\substack{i=1\\i\neq j}}^{n} |a_{i,j}|$$

$$= \|\boldsymbol{r}\|_1 - |r_j| + |r_j| \sum_{\substack{i=1\\i\neq j}}^{n} \left|\frac{a_{i,j}}{a_{j,j}}\right| = \|\boldsymbol{r}\|_1 - |r_j|\delta\,,$$

where $\delta := 1 - \sum_{\substack{i=1\\i\neq j}}^{n} \left|\frac{a_{i,j}}{a_{j,j}}\right| > 0$ because of (1).

If $|r_j| > \sigma_0$ in the manual or threshold relaxation, then $\|\boldsymbol{r}\|_1$ is diminished by more than $\delta\sigma_0$. But since $\|\boldsymbol{r}\|_1$ is bounded from below, this can happen only finitely many times, which concludes the proof.

During cyclic relaxation, $|r_j|$ may be less than σ_0. For this reason the proof of convergence is more difficult, but actually not interesting, because on a computer with a limited machine accuracy *eps* the cyclic relaxation turns into a threshold relaxation anyway.

8.3 The Minimum Problem

A series of problems, of Mathematical Physics and other applied sciences, lead to a quadratic form

$$F(\boldsymbol{p}) := \frac{1}{2}\boldsymbol{p}^{\mathrm{t}} A \boldsymbol{p} - \boldsymbol{p}^{\mathrm{t}}\boldsymbol{a}$$

in $\boldsymbol{p}$ with a symmetric, positive definite matrix A. One has:

Theorem 2: *$F(\boldsymbol{p})$ attains its minimum for the solution $\boldsymbol{p} = \boldsymbol{x}$ of the linear system $A\boldsymbol{x} = \boldsymbol{a}$.*

For the proof let $\boldsymbol{p} = \boldsymbol{x} + \boldsymbol{d}$. Then

$$F(\boldsymbol{p}) = \frac{1}{2}\boldsymbol{d}^{\mathrm{t}} A\boldsymbol{d} + \frac{1}{2}\boldsymbol{x}^{\mathrm{t}} A\boldsymbol{x} - \boldsymbol{x}^{\mathrm{t}}\boldsymbol{a} = F(\boldsymbol{x}) + \frac{1}{2}\boldsymbol{d}^{\mathrm{t}} A\boldsymbol{d} \geq F(\boldsymbol{x})\,,$$

the equality holds only if $\boldsymbol{d} = \mathbf{o}$ because A is positive definite.

Now, the problem is to find the minimum of $F(\boldsymbol{p})$. With the minimum, one gets the solution $\boldsymbol{x}$ of the linear system $A\boldsymbol{x} = \boldsymbol{a}$ as well. To improve an approximation $\boldsymbol{p}$ with the residual $\boldsymbol{r} := A\boldsymbol{p} - \boldsymbol{a}$, one minimizes F in some direction $\boldsymbol{v}$, i.e., one sets

$$\boldsymbol{q} = \boldsymbol{p} + \boldsymbol{v}\lambda\,,$$

whence

$$F(\boldsymbol{q}) = F(\boldsymbol{p}) + \boldsymbol{v}^t\boldsymbol{r}\lambda + \frac{1}{2}\boldsymbol{v}^t A\boldsymbol{v}\lambda^2 .$$

Thus, $F = F(\boldsymbol{q}(\lambda))$ has a relative minimum for

$$\lambda = \lambda' := -\frac{\boldsymbol{v}^t\boldsymbol{r}}{\boldsymbol{v}^t A\boldsymbol{v}} \quad \text{or} \quad \boldsymbol{q} = \boldsymbol{p}' := \boldsymbol{p} + \boldsymbol{v}\lambda'$$

which is

$$F(\boldsymbol{p}') = F(\boldsymbol{p}) - \frac{1}{2}\frac{(\boldsymbol{v}^t\boldsymbol{r})^2}{\boldsymbol{v}^t A\boldsymbol{v}} \leq F(\boldsymbol{p}) .$$

There are several strategies for the choice of $\boldsymbol{v}$:

Coordinate relaxation: For the choice $\boldsymbol{v} = \boldsymbol{e}_j$ one obtains

$$\lambda' = -\frac{r_j}{a_{j,j}} , \quad \boldsymbol{p}' = \boldsymbol{p} - \boldsymbol{e}_j\frac{r_j}{a_{j,j}} , \quad \boldsymbol{r}' = \boldsymbol{r} - \boldsymbol{a}_j\frac{r_j}{a_{j,j}} ,$$

the coordinate relaxation in **8.1**.

Method of steepest descent: It is certainly more effective to progress in the direction of the gradient of F at $\boldsymbol{p}$ rather than in the direction of the maximal $|r_j|$. This direction of the gradient is **locally optimal**. Here, it coincides with the direction $\boldsymbol{r}$. On choosing $\boldsymbol{v} = \boldsymbol{r}$ one gets

$$\lambda' = -\frac{\boldsymbol{r}^t\boldsymbol{r}}{\boldsymbol{r}^t A\boldsymbol{r}} =: -\frac{1}{R}$$

and

$$F(\boldsymbol{p}') = F(\boldsymbol{p}) - \frac{1}{2R}\boldsymbol{r}^t\boldsymbol{r} .$$

This choice of $\boldsymbol{v}$ leads to the following algorithm which is applicable to matrices satisfying the conditions of Theorem 3 given in **8.4**:

Method of Steepest Descent

	Input:	A, pos. def. $n \times n$-matrix; $\boldsymbol{a}, \boldsymbol{p}$ n-columns; $\sigma > 0$ tolerance
	Output:	$\boldsymbol{p}$ with $\|\|A\boldsymbol{p} - \boldsymbol{a}\|\|_2 \leq \sigma$
1	Determine $\boldsymbol{r} := A\boldsymbol{p} - \boldsymbol{a}$	
2	and $\lambda' := -\frac{\boldsymbol{r}^t\boldsymbol{r}}{\boldsymbol{r}^t A\boldsymbol{r}}$.	
3	Set $\boldsymbol{p} := \boldsymbol{p} + \boldsymbol{r}\lambda'$,	
4	and $\boldsymbol{r} := \boldsymbol{r} + A\boldsymbol{r}\lambda'$.	
5	If $\|\|\boldsymbol{r}\|\|_2 \leq \sigma$: **stop** .	
6	Go to **2**.	

Over-relaxation: If $\boldsymbol{v}^t\boldsymbol{r} \neq 0$, then the inequality

$$F(\boldsymbol{p}'') < F(\boldsymbol{p})$$

holds for all $\boldsymbol{p}'' := \boldsymbol{p} + \boldsymbol{v}\omega\lambda'$ with $0 < \omega < 2$. Therefore, $F(\boldsymbol{p})$ is still diminished, when λ' is replaced by $\lambda'' := \omega\lambda'$ in both, the coordinate relaxation or the method of steepest descent, although not to the relative minimum if $\omega \neq 1$. In case $\omega < 1$ this is called **under-relaxation** and in case $\omega > 1$ **over-relaxation.**

Remark 1: There are classes of matrices for which, e.g., the cyclic relaxation converges clearly better with over-relaxation than the ordinary cyclic relaxation with $\omega = 1$.

Remark 2: The quotient $R := \boldsymbol{r}^t A\boldsymbol{r}/\boldsymbol{r}^t\boldsymbol{r}$ is known as the **Rayleigh quotient** of A belonging to $\boldsymbol{r}$. Because of **2.5** holds

$$R \leq \frac{\|\boldsymbol{r}\|_2\|A\boldsymbol{r}\|_2}{\|\boldsymbol{r}\|_2\|\boldsymbol{r}\|_2} = \frac{\|A\boldsymbol{r}\|_2}{\|\boldsymbol{r}\|_2} \leq \|A\|_2 .$$

8.4 Convergence for Symmetric, Positive Definite Matrices

There is an analog to Theorem 1 concerning the convergence for diagonal dominant matrices A.

Theorem 3: *The coordinate relaxation in* **8.1** *and the method of steepest descent converge for the over-relaxation factors* ω *with* $0 < \omega < 2$ *and for symmetric, positive definite matrices* A.

In particular this encompasses the convergence for $\omega = 1$. For the proof consider the attenuation of $F(\boldsymbol{p})$ at the transition from $\boldsymbol{p}$ to $\boldsymbol{p}'' := \boldsymbol{p} + \boldsymbol{e}_j\omega\lambda'$. If $|r_j| > \sigma$ in case of the manual or the threshold relaxation, one gets

$$F(\boldsymbol{p}'') = F(\boldsymbol{p}) - \omega(2-\omega)\frac{1}{2}\frac{r_j^2}{a_{j,j}} < F(\boldsymbol{p}) - \omega(2-\omega)\frac{1}{2}\frac{\sigma^2}{a_{j,j}} .$$

$F(\boldsymbol{p})$ is bounded from below, consequently, $|r_j|$ can be bigger than σ only finitely often.

This is not the case with the cyclic relaxation, but the same considerations as in Theorem 1 apply. For the method of steepest descent one gets

$$F(\boldsymbol{p}'') = F(\boldsymbol{p}) - \omega(2-\omega)\frac{\|\boldsymbol{r}\|_2^2}{2R} < F(\boldsymbol{p}) - \omega(2-\omega)\frac{1}{2}\frac{\sigma^2}{\|A\|_2} .$$

The reasoning is the same as above, with $\|A\|_2$ instead of $a_{j,j}$.

8.5 Geometric Meaning

The minimization of a quadratic form has the following geometric meaning. The level surfaces $F(\boldsymbol{y})$ = constant are similar, concentric ellipsoids. They intersect a plane at similar and concentric ellipsoids, and especially the plane through $\boldsymbol{p}$ containing $\boldsymbol{r}$ and $\boldsymbol{v}$. This is shown in the figure:

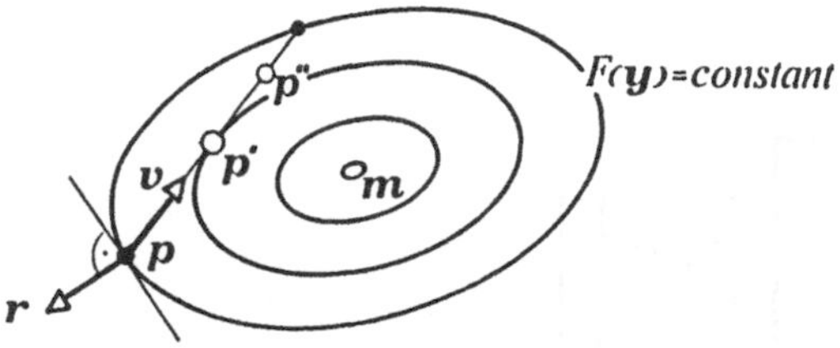

Figure 8.1
Relaxation with symmetric A

A straight line touches one ellipsoid and intersects the others at two points that have the point of contact as their midpoint. From $\boldsymbol{p}$ one progresses in direction $\boldsymbol{v}$ to the point of contact $\boldsymbol{p}'$ (at which $\boldsymbol{r}' \perp \boldsymbol{v}$) or further or not that far if $w \neq 1$, but in any case, one stays within $F(\boldsymbol{y}) = F(\boldsymbol{p})$ because $0 < w < 2$. Clearly, $\boldsymbol{v} = \pm\boldsymbol{r}$ is locally optimal. The common midpoint of all ellipsoids is the solution of the linear system $A\boldsymbol{x} = \boldsymbol{a}$. The the common midpoint of all ellipsoids of the intersection is . These results are proved in Analytic Geometry.

8.6 Notes and Exercises

1. Multiplying the ith equation of the linear system changes the component r_i by the same factor. This can be used to assign special weights to single equations.
2. Two directions $\boldsymbol{v}_i$ and $\boldsymbol{v}_k$ are said to be **conjugate** with respect to the surfaces $F(\boldsymbol{y})$ = constant if $\boldsymbol{v}_i^t A \boldsymbol{v}_k = 0$. If one progresses from $\boldsymbol{p}_j$, $j = 1, \ldots, n$, in direction $\boldsymbol{v}_j$ to the respective minimum $\boldsymbol{p}_{j+1}$ one obtains the exact solution after n steps provided that $\boldsymbol{v}_1 = \boldsymbol{r}_1$ and $\boldsymbol{v}_j$ is conjugate to $\boldsymbol{v}_{j-1}$ with respect to $F(\boldsymbol{y})$ = constant and chosen from the plane spanned by $\boldsymbol{v}_{j-1}$ and $\boldsymbol{r}_j$. (Stiefel and Hestenes 1952).
3. The coordinate relaxation converges for irreducible matrices even if only the **weak column sum criterion** is satisfied:

$$|a_{k,k}| \geq \sum_{\substack{j=1 \\ j \neq k}}^{n} |a_{j,k}|, \quad k = 1, \ldots, n,$$

   where strict inequality must hold for at least one k.

9 Data Fitting

Neither is a linear system always solvable, nor is the solution always unique. A system with more equations than unknowns is generally **overdetermined** while a system with fewer equations than unknowns is generally **underdetermined**.

9.1 Overdetermined Systems of Linear Equations

The linear system $A\boldsymbol{x} = \boldsymbol{a}$

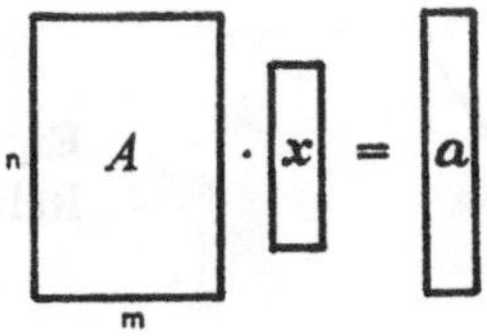

with a "tall" $n \times m$-matrix A, $n > m$, has a solution if and only if rank A = rank $[A, \boldsymbol{a}]$. Otherwise, the residual

$$\boldsymbol{r} := A\boldsymbol{x} - \boldsymbol{a}$$

does not vanish for any $\boldsymbol{x}$. In this case, the **solution** is defined to be some $\boldsymbol{x}$ for which the residual is as small as possible, with respect to some appropriate norm $\|\boldsymbol{r}\|$.

Using the Euclidian norm $\|\boldsymbol{r}\|_2 = \sqrt{\boldsymbol{r}^t\boldsymbol{r}}$ leads to the **method of least squares** already studied by Gauss. Assuming maximal rank m for A one can prove the following theorem:

Theorem 1: *$\|\boldsymbol{r}\|_2$ is minimal for $\boldsymbol{x}$ with $A^t\boldsymbol{r} = \mathbf{o}$.*

For the proof let $\boldsymbol{r}'$ be the residual belonging to $\boldsymbol{x}' := \boldsymbol{x} + \boldsymbol{d}$, i.e.,

$$\boldsymbol{r}' := \boldsymbol{r} + A\boldsymbol{d}.$$

Then, $A^t\boldsymbol{r} = \mathbf{o}$ implies

$$\|\boldsymbol{r}'\|_2^2 = \boldsymbol{r}^t\boldsymbol{r} + 2\boldsymbol{d}^tA^t\boldsymbol{r} + \boldsymbol{d}^tA^tA\boldsymbol{d} = \|\boldsymbol{r}\|_2^2 + \|A\boldsymbol{d}\|_2^2.$$

This means $\|\boldsymbol{r}'\|_2 > \|\boldsymbol{r}\|_2$ holds for all $\boldsymbol{d} \neq \mathbf{o}$ and therefore for all $\boldsymbol{r}' \neq \boldsymbol{r}$.

Inserting $\boldsymbol{r}$ into $A^t\boldsymbol{r} = \mathbf{o}$ yields

$$A^tA\boldsymbol{x} - A^t\boldsymbol{a} = \mathbf{o}$$

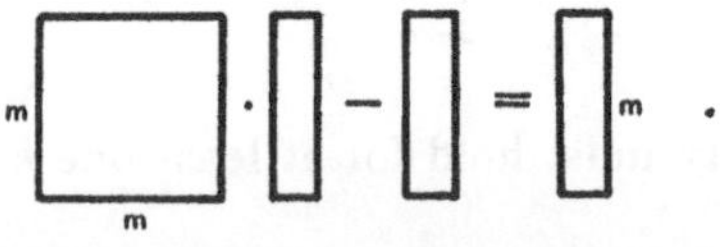

These equations are called **normal equations** since they express the fact that the residual $Ax - a$ should be normal or perpendicular to the columns of A. Their matrix is symmetric and positive definite because

$$[A^t A]^t = [A]^t [A^t]^t = A^t A,$$

and

$$y^t A^t A y = [Ay]^t [Ay] > 0 \text{ for all } y \neq o.$$

Remark 1: The normal equations can be solved by a Cholesky factorization of $A^t A$. Yet, it is better to use a direct QU factorization of A, since they are generally ill-conditioned.

9.2 Using the QU Factorization

The $n \times m$-matrix A with m linearly independent columns requires m Householder reflections H_j to transform the matrix $[A, a]$ into a matrix U consisting of the first m columns of a triangular matrix and a column b:

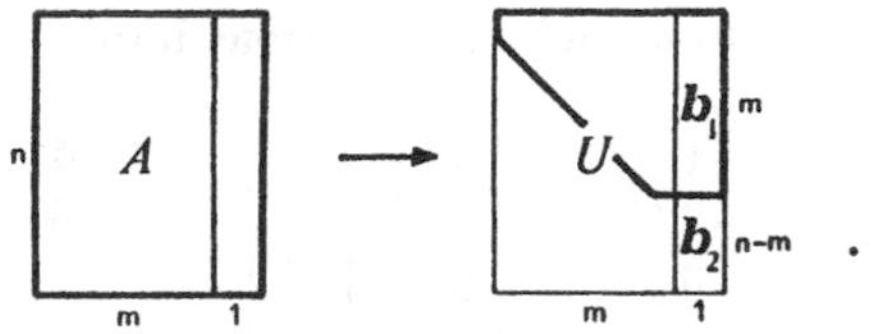

The Euclidean norm of any column is preserved since the H_j are orthonormal. This means particularly for the residual

$$||r||_2 := ||Ax - a||_2 = ||Ux - b||_2 .$$

Hence, $||r||_2$ becomes minimal for

$$U_1 x = b_1 .$$

This minimal value is $||r||_2 = ||b_2||_2$, with $[U_1, b_1]$ the first m rows of $[U, b]$ and b_2 the lower $n - m$ entries of b.

9.3 Application

Overdetermined linear systems occur, for instance, with the problem of finding the linear combination

$$y = f(t) := \sum_{k=1}^{m} c_k f_k(t)$$

of m linearly independent functions which approximates n given values y_i best at n distinct fixed numbers t_i. Inserting the n pairs t_i, y_i into $y = f(t)$ yields a system of linear equations for the c_k which is usually overdetermined in case $n > m$. Its matrix is $[f_k(t_i)]$.

Example 1: The four points of the table

$t =$	-1	0	1	2
$y =$	2	1	2	3

are to be fitted by the parabola

$$y = c_0 + c_1 t + c_2 t^2 .$$

Inserting the points into the equation of the parabola gives the linear system

$$\begin{bmatrix} 1 & -1 & 1 \\ 1 & 0 & 0 \\ 1 & 1 & 1 \\ 1 & 2 & 4 \end{bmatrix} \begin{bmatrix} c_0 \\ c_1 \\ c_2 \end{bmatrix} = \begin{bmatrix} 2 \\ 1 \\ 2 \\ 3 \end{bmatrix} .$$

It is transformed by three Householder reflections into

$$\begin{bmatrix} -2 & -1 & -2.99 \\ 0 & -2.24 & -2.25 \\ 0 & 0 & 1.99 \\ 0 & 0 & 0 \end{bmatrix} \begin{bmatrix} c_0 \\ c_1 \\ c_2 \end{bmatrix} = \begin{bmatrix} -4 \\ -0.89 \\ 1.01 \\ 0.45 \end{bmatrix} .$$

The solution $\begin{bmatrix} c_0 \\ c_1 \\ c_2 \end{bmatrix} = \begin{bmatrix} 1.3 \\ -0.1 \\ 0.5 \end{bmatrix}$ minimizes $\boldsymbol{r}$, giving $\begin{bmatrix} r_1 \\ r_2 \\ r_3 \\ r_4 \end{bmatrix} = \begin{bmatrix} -0.1 \\ 0.3 \\ -0.3 \\ 0.1 \end{bmatrix}$, and $||\boldsymbol{r}||_2^2 = 0.2$. The corrected ordinates are $y_i + r_i$.

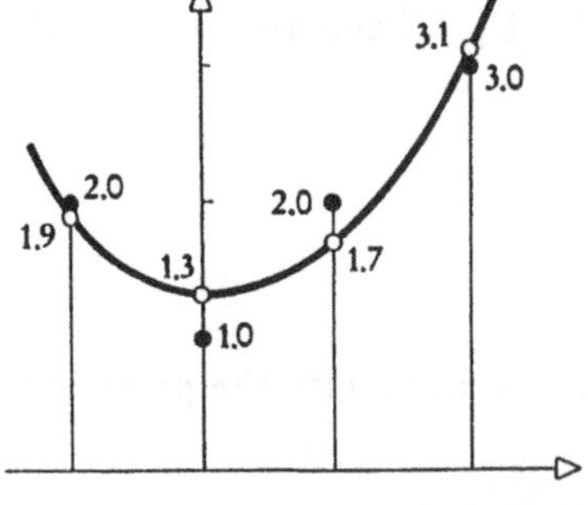

Figure 9.1
Least-squares parabola

9.4 Underdetermined Systems of Linear Equations

The linear system

(2) $$By = b$$

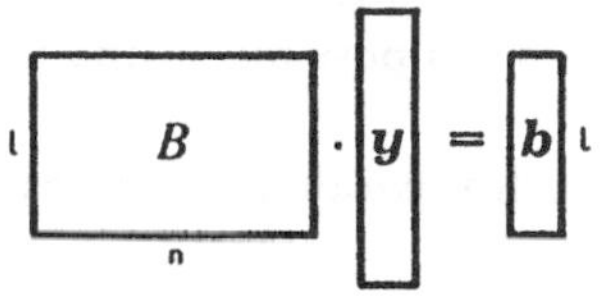

with a "wide" $l \times n$-matrix, $n > l$, has solutions only if rank B = rank $[B, b]$. Among all solutions one is usually interested in one that lies closest to a given p, where the distance $||y - p||$ is measured by some appropriate norm.

Using the Euclidean norm $||y - p||_2$ leads again to a solution already employed by Gauss in land surveying. If B has maximal rank l, one can prove:

Theorem 2: *$||y - p||_2$ is minimal for $y - p = B^t t$.*

For the proof let $y' := y + d$ be a second solution of (2) besides y. This implies $Bd = o$ and

$$||y' - p||_2^2 = [y - p]^t[y - p] + 2d^t[y - p] + d^t d,$$

and because of

$$d^t[y - p] = d^t B^t t = o^t t = 0$$

also

$$||y' - p||_2^2 > ||y - p||_2^2 \text{ for all } d \neq o,$$

or all $y' \neq y$ respectively.

Multiplying $y - p = B^t t$ by B and using (2) leads to

$$BB^t t = b - Bp$$

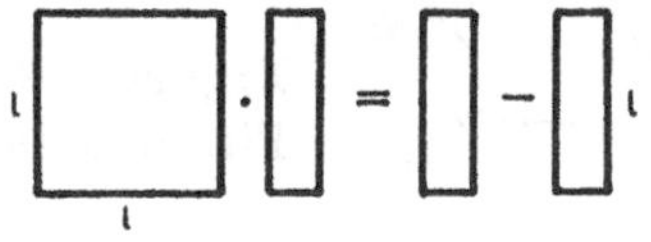

These equations are also called **normal equations** after Gauss. Their matrix is again symmetric and positive definite.

The **correlate equations**

$$y = p + B^t t$$

together with the solution t of the normal equations determine y.

Remark 2: The normal equations can be solved by a Cholesky factorization. However, since they are ill-conditioned it is advisable to use a direct QU decomposition of B^t.

9.5 Application

Underdetermined systems of linear equations occur, for instance, if the data p_i must satisfy linear conditions for theoretical reasons, like the currents in a network or the angles of a triangulation. In those cases one wants values y_i which satisfy these conditions and are as close as possible to the p_i.

Example 2: The values y_i which a parabola assumes at four equidistant abscissae satisfy

$$y_1 - 3y_2 + 3y_3 - y_4 = 0$$

according to **22.1**. Consider the problem of finding the y_i that are close to the values

$$p_i = 2,\ 1,\ 2,\ 3\,.$$

It is $B = [1, -3, 3, -1]$, $\boldsymbol{b} = \mathbf{o}$ and consequently $BB^t = 20$. $\boldsymbol{t}$ has only one component $t = -0.1$. Thus

$$\boldsymbol{y} = \boldsymbol{p} + B^t\boldsymbol{t} = [1.9,\ 1.3,\ 1.7,\ 3.1]^t\,.$$

The parabola is depicted in Figure 9.1.

9.6 Geometric Meaning and Duality

Theorems 1 and 2 allow the following geometric interpretation in the Euclidean space $\mathbb{E}^n$:

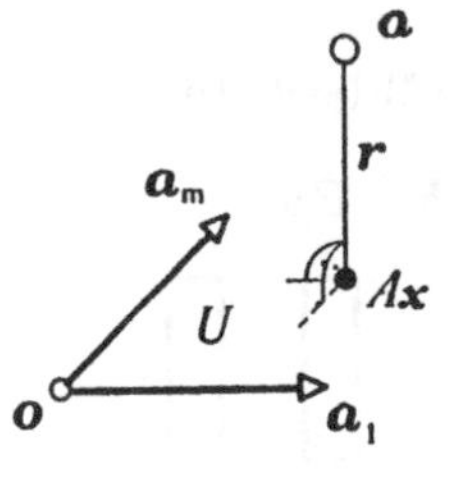

Figure 9.2
Fitting data

$A^t\boldsymbol{r} = \mathbf{o}$ means: $A\boldsymbol{x} = \boldsymbol{a} + \boldsymbol{r}$ is the foot of the perpendicular from $\boldsymbol{a}$ onto the subspace U with the parametric representation $\boldsymbol{y} = A\boldsymbol{x}$.

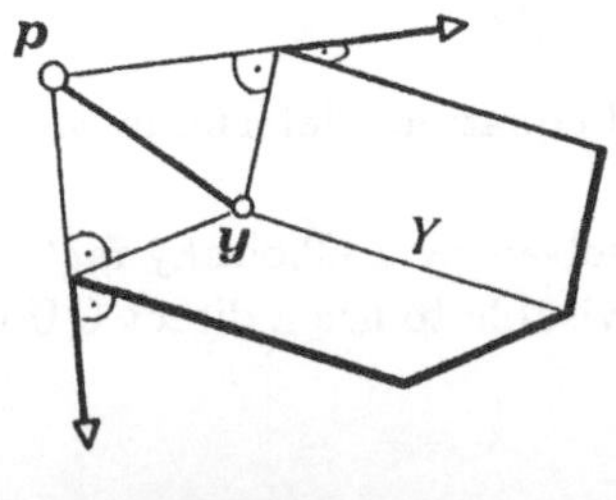

Figure 9.3
Fitting data under constraints

$\boldsymbol{y}-\boldsymbol{p}=B^{t}\boldsymbol{t}$ means: $\boldsymbol{y}$ is the foot of the perpendicular from $\boldsymbol{p}$ onto the subspace Y with the implicit representation $B\boldsymbol{y}=\boldsymbol{b}$.

In both cases one has to drop the perpendicular from a point onto a subspace. Both problems differ only in the representation of the subspaces. It is a subject of Analytic Geometry to transform one into the other, the **dual form**.

Two fitting problems (1) and (2), for instance, are directly dual to each other if

$$m+l=n\,,\quad \boldsymbol{b}=\mathbf{o}\,,\quad BA=O \text{ and } \boldsymbol{p}=\boldsymbol{a}\,.$$

Example 3: The two examples 1 and 2 are dual versions of the same problem.

9.7 Notes and Exercises

1. The normal equations for Example 1 are

$$\begin{bmatrix} 4 & 2 & 6 \\ 2 & 6 & 8 \\ 6 & 8 & 18 \end{bmatrix} \begin{bmatrix} c_0 \\ c_1 \\ c_2 \end{bmatrix} = \begin{bmatrix} 8 \\ 6 \\ 16 \end{bmatrix}.$$

2. The Cholesky factorization of the normal equations of Example 1 yields

$$\begin{bmatrix} 2 & 1 & 3 \\ 0 & \sqrt{5} & \sqrt{5} \\ 0 & 0 & 2 \end{bmatrix} \begin{bmatrix} c_0 \\ c_1 \\ c_2 \end{bmatrix} = \begin{bmatrix} 4 \\ 0.4\sqrt{5} \\ 1 \end{bmatrix}.$$

3. Chebyshev considered the maximum norm in minimizing the residual. This results in systems of linear inequalities which are the topic of **10**.

4. If $A\boldsymbol{x}-\boldsymbol{a}=\boldsymbol{r}=\mathbf{o}$ is "almost" solvable, i.e., if $\|\boldsymbol{r}\|_2 \ll \|\boldsymbol{a}\|_2$ for the "solution" $\boldsymbol{x}$ of the normal equations, then relative changes of $\boldsymbol{a}$ and A are basically transferred over to $\boldsymbol{x}$ with the factor cond A and it is advisable to use the QU factorization of A.

5. If the residual $\boldsymbol{r}=A\boldsymbol{x}-\boldsymbol{a}$ of the "solution" $\boldsymbol{x}$ is "large" in comparison with $\boldsymbol{a}$, then relative changes of $\boldsymbol{a}$ and A are transferred with the factor cond $A^{t}A$. In this case the solution of the data fitting problem obtained from the normal equations will be approximately as precise as the solution obtained from the QU factorization of A. For this reason the QU factorization is always the better choice.

10   Linear Optimization

Frequently in practice the problem arises to maximize a function of m variables x_k which are subject to certain constraints. The simplest case is that of **linear optimization**.

10.1   Linear Inequalities and Linear Programming

The $x \in \mathbb{R}^m$ satisfying an inequality

$$y = a^t x + a \geq 0$$

form a **half-space** of $\mathbb{R}^m$ which is bounded by the **hyperplane** $y = 0$. In case $a \geq 0$, $x = o$ lies in this half-space.

The $x \in \mathbb{R}^m$ satisfying a system of l linear inequalities[1]

(1) $y = Ax + a \geq 0,$

$$l \begin{vmatrix} y \end{vmatrix} = \begin{vmatrix} & A & \\ & & \\ & m & \end{vmatrix} \begin{vmatrix} x \end{vmatrix} + \begin{vmatrix} a \end{vmatrix} \geq \begin{vmatrix} o \end{vmatrix} l$$

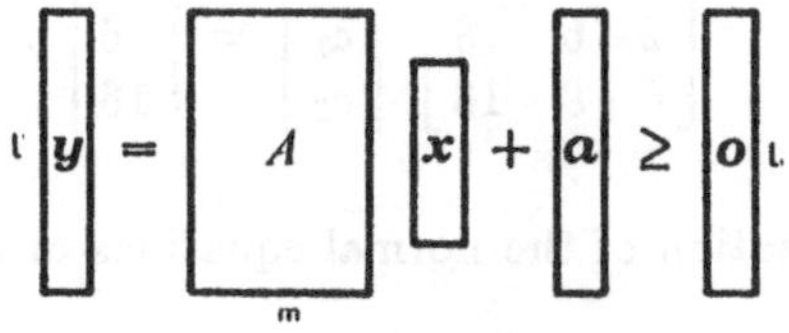

with the $l \times m$-matrix A and the l-column a, form a **simplex** S. This implies immediately: The intersection of two simplices is a simplex and therefore: A simplex is **convex**, i.e., S contains for any two of its points s, t all the other points of the line $[s, t]$ as well. Let $l > m$.

A point of the simplex S, for which at least m linearly independent y_i vanish is called **vertex** of the simplex. A vertex is **simple** if precisely m linearly independent y_i vanish there. Introducing m such y_i of a simple vertex as new coordinates puts (1) in the **standard form**

(2)
$$y_1 \geq 0,$$
$$y_2 = B y_1 + b \geq 0.$$

$$m \begin{vmatrix} y_1 \end{vmatrix} \geq \begin{vmatrix} o \end{vmatrix} , \quad n \begin{vmatrix} y_2 \end{vmatrix} = \begin{vmatrix} & B & \\ & m & \end{vmatrix} \cdot \begin{vmatrix} y_1 \end{vmatrix} - \begin{vmatrix} b \end{vmatrix} \geq \begin{vmatrix} o \end{vmatrix} n .$$

---

[1] By $a \geq b$ it is meant $a_i \geq b_i$ for all i.

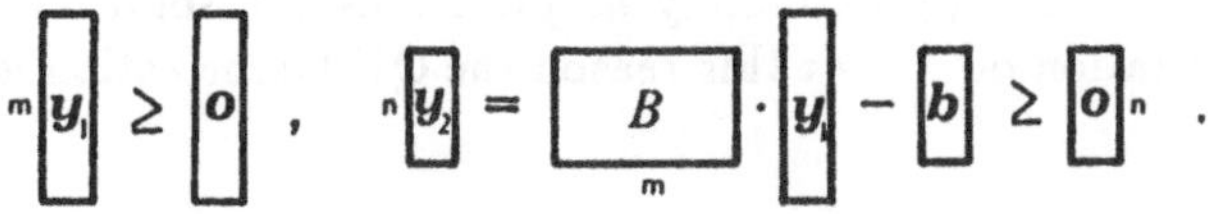

where y_1 consists of the m y_i, and y_1 and y_2 decompose y in an m- and an n-column with $n := l - m$. Further, B is an $n \times m$-matrix and b an n-column. Because $y_1 = o$ lies in S and is simple, $b \geq o$.

The task of a **linear programming problem** in $\mathbb{R}^m$ is to maximize a linear **objective function**

(3) $$z := z^t x = c^t y_1 + c$$

subject to the l linear constraints (1) or (2), i.e., to search a point of the simplex at which z assumes its maximum in S. The points of S are called **feasible** in this context.

A solution can be obtained quite simply. First (1) is transformed into the **standard form** (2), then the components of y_1 are exchanged with appropriate components of y_2 until z cannot be further increased for any component of y_1. The exchange algorithm **5.1** serves to interchange the variables in the scheme

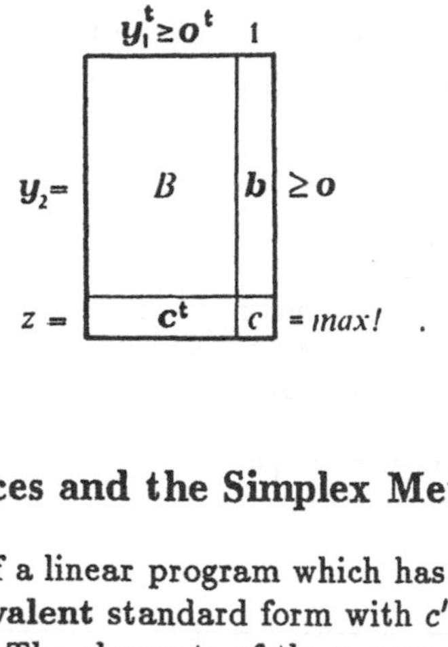

10.2 Exchanging Vertices and the Simplex Method

The standard form (2), (3) of a linear program which has only simple vertices can be transformed into an **equivalent** standard form with $c' > c$ by exchanging some y_s of y_1 with some y_r of y_2. The elements of the program after the exchange are denoted by $'$.

According to **5.1** the point $y_1' = o$ lies in S iff

$$b_r' := \frac{-b_r}{b_{r,s}} > 0 \quad \text{and} \quad b_i' := b_i - \frac{b_{i,s}}{b_{r,s}} b_r > 0\,, \quad i \neq r\,.$$

Also, z is bigger at $y_1' = o$ than at $y_1 = o$ iff

$$c' := c - \frac{b_r c_s}{b_{r,s}} > c\,.$$

These three conditions are equivalent to

$$c_s > 0\,, \qquad b_{r,s} < 0\,, \qquad \frac{b_r}{b_{r,s}} = \max_{b_{i,s} < 0} \frac{b_i}{b_{i,s}}\,.$$

The following things may happen in exchanging two vertices. If there is a pivot $b_{r,s}$ fulfilling the conditions, one can exchange y_s and y_r and obtain a standard form equivalent to (2), (3).

However, if there is no $c_s > 0$, then $z \leq z(\mathbf{o})$ holds for all $\mathbf{y}_1 \geq \mathbf{o}$, i.e., z is maximal at $\mathbf{y}_1 = \mathbf{o}$ with $z = c$. Thus, $\mathbf{y}_1 = \mathbf{o}$ with $\mathbf{y}_2 = \mathbf{b}$ is a solution of the problem. Obviously, if $c_s = 0$, then $z = c$ in the direction of y_s, etc.

In case there is no $b_{r,s} < 0$ for some $c_s > 0$, then the objective function can grow arbitrary large within the simplex because, then, $\mathbf{y}_2 > \mathbf{o}$ for all points on the y_s-axis with $y_s \leq 0$; **no solution** exists. One continues to exchange vertices until no further pivot can be found. This algorithm terminates since the simplex has at most $\binom{n+m}{m}$ vertices and none of them is met twice because z increases strictly with each exchange.

Simplex

Input: normal form (2), (3) of a linear programming problem

Output: normal form with $z = \max$ for $\mathbf{y}_1 = \mathbf{o}$

Search $c_s > 0$.

If all $c_k \leq 0$: **stop** .

Determine r with $\dfrac{b_r}{b_{r,s}} = \max\limits_{b_{i,s}<0} \dfrac{b_i}{b_{i,s}}$,

if all $b_{i,s} \geq 0$: **stop**, no solution.

Execute **Exchange r, s** and

save r, s .

Go to **1** .

Example 1: Solving a linear programming problem in standard form is demonstrated with the scheme below. The pivot rows and columns are marked. The **quotients** $b_i/b_{i,s}$ for $b_{i,s} < 0$ are written on the right side of the scheme:

	y_1	y_2	1			
$y_3 =$	1	−2	4	≤ 0	$\frac{4}{2}$	◄
$y_4 =$	−1	−1	8	≤ 0	$\frac{8}{1}$	
$z =$	1	4	0			
		▲				

Exchanging y_3 and y_2 converts the scheme into

	y_1	y_3	1		
$y_2 =$	$\frac{1}{2}$	$-\frac{1}{2}$	2	≥ 0	
$y_4 =$	$-\frac{3}{2}$	$\frac{1}{2}$	6	≥ 0	$\frac{12}{3}$ ◀
$z =$	3	-2	8		
	▲				

,

and the subsequent exchange of y_1 and y_4 converts it into

	y_4	y_3	1	
$y_2 =$	$-\frac{1}{3}$	$-\frac{1}{3}$	4	≥ 0
$y_1 =$	$-\frac{2}{3}$	$\frac{1}{3}$	4	≥ 0
$z =$	-2	-1	20	

Since there is no $c_s > 0$ anymore, the linear programming problem has the solution

$$y_4 = 0\,, \quad y_3 = 0 \text{ with } y_2 = 4\,, \quad y_1 = 4\,.$$

There, z assumes its maximum $c = 20$.

The figure shows the corresponding simplex:

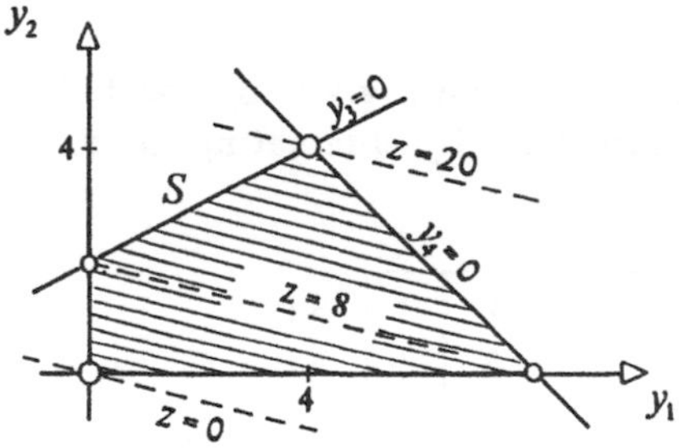

Figure 10.1
Simplex method

10.3 Elimination

The elimination of the x_k to transform a generic linear programming problem (1), (3) into an equivalent standard form (2), (3) is even simpler than the simplex method itself.

First one chooses a feasible point as the origin. Then $\boldsymbol{x} = \mathbf{o}$ is feasible and $\boldsymbol{a} \geq \mathbf{o}$.

Secondly, the x_s are successively exchanged, $s = 1, \ldots, m$, with some suitable y_r. It is expedient here, to transpose the rows r and s of A and to rename the variables y_r and y_s by y_s and y_r respectively after each exchange.

According to **5.1** the following condition has to be satisfied in order to maintain the property that the a_i, $i = s+1, \ldots, n$, stay non-negative. For the exchange of x_s with some y_r, $s+1 \leq r \leq n$, one must have

$$a_i - \frac{a_{i,s}}{a_{r,s}} a_r \geq 0, \quad i = s, \ldots, n \text{ but } i \neq r.$$

Clearly, this condition needs to be checked only if $a_{i,s}/a_{r,s} > 0$. In general for each s, there are two pivots $a_{r,s}$ satisfying these conditions. Eventually, after all x_s are exchanged, the general problem (1), (3) splits into the representation of x

$$x = Py_1 + p,$$

with an $m \times m$-matrix P and a (not necessarily positive) m-column p, and the standard form (2), (3):

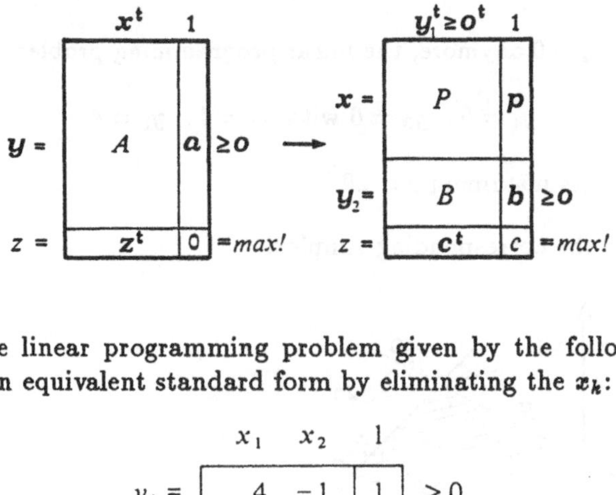

Example 2: The linear programming problem given by the following scheme is converted into an equivalent standard form by eliminating the x_k:

	x_1	x_2	1	
$y_1 =$	4	-1	1	≥ 0
$y_2 =$	10	-2	1	≥ 0
$y_3 =$	-4	1	1	≥ 0
$y_4 =$	-10	2	1	≥ 0
$z =$	1	0	0	$=$ max! .

a is already positive and one may exchange x_1 with y_4 and then x_2 with y_1. This splits the scheme into

	y_4	y_1	1
$x_2 =$	-2	-5	7
$x_1 =$	-0.5	-1	1.5

,

and

	y_4	y_1	1	
$y_2 =$	-1	0	2	≥ 0
$y_3 =$	0	-1	2	≥ 0
$z =$	-0.5	-1	1.5	$= \max!$.

Since no $c_s > 0$, one cannot further increase z. Hence, the linear programming problem has the solution

$$y_1 = y_4 = 0 \text{ with } y_2 = 2\,,\; y_3 = 2\,,\; x_2 = 7\,,\; x_1 = 1.5 \text{ and } \max z = 1.5\,.$$

10.4 Data Fitting after Chebyshev

Linear programming problems arise also in solving overdetermined systems of linear equations[1)]

$$(4) \qquad A\boldsymbol{x} - \boldsymbol{a} = \boldsymbol{r} \neq \mathbf{o}$$

after Chebyshev, i.e., in minimizing the maximum norm

$$r := \|\boldsymbol{r}\|_\infty := \max_i |r_i|$$

of the residual. This optimization problem can be put in the form (1), (3). Since $r > 0$, new coordinates

$$x_0 := \frac{1}{r}\,, \qquad \bar{x}_i := \frac{1}{r} x_i$$

can be introduced. This translates (4) into

$$-\boldsymbol{a}x_0 + A\bar{\boldsymbol{x}} = \boldsymbol{r}x_0\,.$$

Because the normalized residual satisfies the inequalities

$$\pm \boldsymbol{r}\frac{1}{r} \leq \boldsymbol{e} := [1, \ldots, 1]^t\,,$$

the optimization problem can finally be written as

$$\begin{aligned} +\boldsymbol{a}x_0 - A\bar{\boldsymbol{x}} + \boldsymbol{e} &\geq \mathbf{o} \\ -\boldsymbol{a}x_0 + A\bar{\boldsymbol{x}} + \boldsymbol{e} &\geq \mathbf{o} \\ z = x_0 &= \max ! \end{aligned}$$

1) The notation complies with 9.

This is a linear programming problem of the form (1), (3). In particular, the point $x_0 = 0$, $\bar{x} = o$ lies in S. So, the problem can be instantly transformed into standard form by eliminations of the x_0, $\bar{x}$, and be solved by the algorithm **simplex**. The solution, finally, has to be converted to the original coordinates:

$$x = \bar{x}\frac{1}{x_0} \text{ and } r = \frac{1}{x_0}.$$

Example 3: Consider the simple overdetermined linear system

$$x - 4 = r_1$$
$$2x - 10 = r_2.$$

The corresponding linear programming problem in non-standard form is given by the scheme

	x_0	$\bar{x}_1$	1	
$y_1 =$	4	−1	1	≥ 0
$y_2 =$	10	−2	1	≥ 0
$y_3 =$	−4	1	1	≥ 0
$y_4 =$	−10	2	1	≥ 0
$z =$	1	0	0	= max! .

It was already solved with different labels in Example 2. Making use of that result, i.e.,

$$x_0 = 1.5 \quad \text{and} \quad \bar{x}_1 = 7,$$

the solution here becomes

$$x = \frac{7}{1.5} = 4.67 \text{ with } r = \frac{1}{1.5} = 0.67.$$

10.5 Notes and Exercises

1. The elimination of x and the algorithm **Simplex** have the following geometric interpretation. First one searches a vertex of the simplex and from there, a sequence of connected edges on which z increases until it reaches its maximum.
2. If there are vertices lying on more than m hyperplanes one must permit $b_i \geq 0$.
3. Linear equations

$$a^t x + a = 0$$

can be written as pairs of inequalities

$$-\boldsymbol{a}^{t}\boldsymbol{x} - a \geq 0\,,$$
$$+\boldsymbol{a}^{t}\boldsymbol{x} + a \geq 0\,.$$

4. The interchange of v_s and u_r in

$$-\boldsymbol{v} = A^{t}\boldsymbol{u}$$

(without exchanging the negative sign) and the interchange of x_s and y_r in

$$\boldsymbol{y} = A\boldsymbol{x}$$

are affected by the same algorithm on the $a_{i,k}$ (duality of the exchange algorithm).

5. The algorithm **Simplex** for the linear programming problem

$$\begin{aligned} \boldsymbol{y}_1 &\geq \mathbf{o} \\ \boldsymbol{y}_2 = \boldsymbol{B}\boldsymbol{y}_1 + \boldsymbol{b} &\geq \mathbf{o} \\ z = \mathbf{c}^{t}\boldsymbol{y}_1 + c &= \max\,! \end{aligned}$$

also solves the dual problem

$$\begin{aligned} \boldsymbol{v}_2 &\leq \mathbf{o} \\ \boldsymbol{v}_1 = -B^{t}\boldsymbol{v}_2 + \mathbf{c} &\leq \mathbf{o} \\ w = -\boldsymbol{b}^{t}\boldsymbol{v}_2 + c &= \min\,!\,. \end{aligned}$$

Both schemes differ just in their labelling:

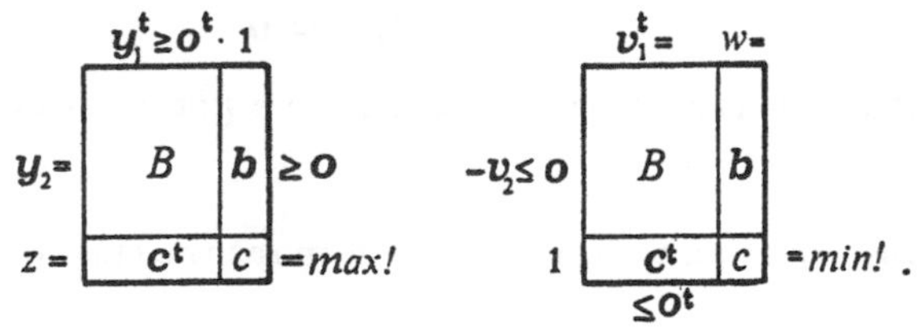

Only the geometric interpretation is different.

6. The standard form of a linear programming problem (2), (3) can be written also as

$$\begin{aligned} -\boldsymbol{y}_1 &\leq \mathbf{o} \\ -\boldsymbol{y}_2 = -\boldsymbol{B}\boldsymbol{y}_1 - \boldsymbol{b} &\leq \mathbf{o} \\ -z = -\mathbf{c}^{t}\boldsymbol{y}_1 - c &= \min\,! \end{aligned}$$

and be solved in dual form.

III Iteration

There are problems that cannot be solved directly. In many cases one depends on the improvement of an approximate solution. The repeated refinement of such an approximation is called **iteration**. The relaxation methods in section **8** are examples for such an iterative solving of a problem.

11 Vector Iteration

Some problems have solutions only for certain parameter values which are called "eigenvalues of the problem". A frequently occurring task in this context is to determine the eigenvalues and eigenvectors of a matrix.

11.1 The Eigenvalue Problem for Matrices

The homogeneous linear system

$$[A - \lambda I]\mathbf{x} = \mathbf{o} \tag{1}$$

with the $n \times n$-matrix A has non-trivial solutions $\mathbf{x} \neq \mathbf{o}$ if and only if

$$\det[A - \lambda I] = 0 \,.$$

This equation is the **characteristic equation** of A. The left side is a polynomial of degree n in λ. Its roots $\lambda_1, \ldots, \lambda_n$ are called **eigenvalues** of A. Only for these λ_i the linear system

$$[A - \lambda_i I]\mathbf{x}_i = \mathbf{o}$$

has non-trivial solutions $\mathbf{x}_i$ which are called the **eigenvectors** of A associated with λ_i.

The linear transformation associated with A maps the eigenvectors onto a multiple of themselves:

$$A\mathbf{x}_i = \mathbf{x}_i \lambda_i \,.$$

This fact already implies that the eigenvalues of a linear transformation do not depend on the basis. Namely, introducing new coordinates $\bar{\mathbf{x}}$ by some non-singular $n \times n$-matrix B, i.e., $\mathbf{x} = B\bar{\mathbf{x}}$, changes (1) into the equation

$$B^{-1}[A - \lambda I]B\bar{\mathbf{x}} = \mathbf{o}$$

which is equivalent to

$$[B^{-1}AB - \lambda I]\bar{\mathbf{x}} = \mathbf{o} \,.$$

The matrices A and $\bar{A} := B^{-1}AB$ are called **similar**. A and $\bar{A}$ possess the same eigenvalues λ_i because

$$\det B^{-1}[A - \lambda I]B = \det B^{-1} \det B \det[A - \lambda I] = \det[A - \lambda I] .$$

The relation between the corresponding eigenvectors is $\boldsymbol{x}_i = B\bar{\boldsymbol{x}}_i$.

Remark 1: Even real matrices can have complex eigenvalues and eigenvectors. In this case they occur in conjugate complex pairs.

11.2 Vector Iteration after von Mises

If λ_j is a p-fold root of the characteristic equation of A, then, A has according to a theorem of linear algebra, at most p linearly independent eigenvectors associated with λ_j. Here it shall always be assumed that A has this maximum number of linearly independent eigenvectors. Consequently, A has n linearly independent eigenvectors $\boldsymbol{x}_i$ associated with its n (not necessarily distinct) eigenvalues λ_i.

The non-singular $n \times n$-matrix comprising these eigenvectors,

$$X := [\boldsymbol{x}_1, \ldots, \boldsymbol{x}_n] ,$$

in some order is called **modal matrix**. It satisfies the equation

$$AX = X\Lambda$$

with the diagonal matrix

$$\Lambda := \begin{bmatrix} \lambda_1 & & 0 \\ & \ddots & \\ 0 & & \lambda_n \end{bmatrix}$$

of the n eigenvalues.

Obviously, one can compute the eigenvalues of A from its characteristic polynomial. However, this is not advisable since the eigenvalues depend far more sensitively on the coefficients of the characteristic polynomial than on the entries of A. Therefore, it is better to calculate the eigenvalues "directly". The classical method of such a direct computation is that of von Mises. It is called the **power method**.

A sequence of vectors is formed with some **starting vector** $\boldsymbol{y}_0 \neq \boldsymbol{o}$ by

$$\boldsymbol{y}_k := A\boldsymbol{y}_{k-1} , \quad k = 1, 2, \ldots .$$

From time to time the vectors have to be normalized

$$\boldsymbol{y}_k := \boldsymbol{y}_k \frac{1}{||\boldsymbol{y}_k||}$$

with some suitable norm $\|\cdot\|$. The following result applies.

Theorem 1: *If A has exactly one eigenvalue λ_1 of maximum modulus and if y_0 is a suitable starting vector, then the sequence*

$$y_k \cdot \frac{1}{\|y_k\|}$$

converges to the normalized eigenvector x_1 associated with λ_1.

The proof is especially simple if there are n linearly independent eigenvectors, which is assumed here; so y_0 has the unique representation

$$y_0 = Xc \text{ with } c := [c_1, \ldots, c_n]^t .$$

From this one can derive

$$y_1 = Ay_0 = AXc = X\Lambda c = X\Lambda_1 c\lambda_1$$

with

$$\Lambda_1 := \frac{1}{\lambda_1}\Lambda = \begin{bmatrix} 1 & & & 0 \\ & \frac{\lambda_2}{\lambda_1} & & \\ & & \ddots & \\ 0 & & & \frac{\lambda_n}{\lambda_1} \end{bmatrix} .$$

After k steps one has

$$\frac{y_k}{\|y_k\|} = \frac{X\Lambda_1^k c}{\|X\Lambda_1^k c\|} .$$

Because $|\lambda_i/\lambda_1| < 1$ for $i \neq 1$, the sequence Λ_1^k converges to the matrix $[e_1, o, \ldots, o]$ which has a 1 in the upper left corner and otherwise zeros. Therefore $X\Lambda_1^k c$ converges to $x_1 c_1$ and $y_k/\|y_k\|$ to the normalized eigenvector x_1 associated with the dominant eigenvalue λ_1 provided that $c_1 \neq 0$.

It follows immediately that for large N

$$y_N := Ay_{N-1} \approx y_{N-1}\lambda_1 .$$

Consequently, one has

Corollary 1:

$$\lim_{N\to\infty} \frac{\|y_N\|}{\|y_{N-1}\|} = |\lambda_1| .$$

Corollary 2:

$$\lim_{N\to\infty} \frac{y_N^t y_N}{y_N^t y_{N-1}} = \lambda_1 .$$

The last quotient is known as the **Schwarz quotient.**

An algorithm is given below. It uses the maximum norm since it is easy to determine:

Power Method

Input: A, $n \times n$-matrix; $\boldsymbol{y}_0$ n-column;
N maximal number of iterations; $\varepsilon > 0$ tolerance

Output: λ_1 eigenvalue; $|\lambda_1| > |\lambda_i|$ for $i \neq 1$;
$\boldsymbol{x}_1$ eigenvector associated with λ_1

1 Set $\nu_0 := \|\boldsymbol{y}_0\|_\infty$.
2 For $k = 1, 2, \ldots, N$
3 set $\boldsymbol{y}_{k-1} := \boldsymbol{y}_{k-1} \dfrac{1}{\nu_{k-1}}$.
4 Determine $\boldsymbol{y}_k := A\boldsymbol{y}_{k-1}$
5 and $\nu_k := \|\boldsymbol{y}_k\|_\infty$.
6 If $\|\boldsymbol{y}_k (\mp) \boldsymbol{y}_{k-1}\nu_k\|_\infty \leq \varepsilon$: go to **7**.
7 Set $\boldsymbol{x}_1 := \boldsymbol{y}_k \dfrac{1}{\nu_k}$
8 and $\lambda_1 := \dfrac{\boldsymbol{y}_k^t \boldsymbol{y}_k}{\boldsymbol{y}_k^t \boldsymbol{y}_{k-1}}$.

In case $k = N$, the algorithm ends in general without $\boldsymbol{y}_N \approx \boldsymbol{y}_{N-1}\lambda_1$.

Remark 2: In practice, one can hardly examine whether $c_1 \neq 0$. However, this is not very important, since the opposite $c_1 = 0$ can neither be presented exactly in a computer nor be maintained because of round-off errors, anyway.

11.3 Inverse Iteration

If A is non-singular, all eigenvalues are distinct from zero and A^{-1} exists. Then $A\boldsymbol{x}_i = \boldsymbol{x}_i\lambda_i$ implies

$$\boldsymbol{x}_i\lambda_i^{-1} = A^{-1}\boldsymbol{x}_i \,.$$

This means any eigenvector $\boldsymbol{x}_i$ of A associated with the eigenvalue λ_i is also an eigenvector of A^{-1} associated with the eigenvalue λ_i^{-1}.

If, in particular, A possesses exactly one eigenvalue λ_n of minimal absolute value, then λ_n^{-1} is the unique eigenvalue of A^{-1} with maximum absolute value, hence it can be calculated as in **11.2**.

To this end one forms the sequence

$$(2) \qquad y_k := A^{-1} y_{k-1}\,, \quad k = 1, 2, \ldots,$$

where y_k is normalized occasionally. However, it is numerically more advantageous to replace (2) by the linear equation

$$A y_k = y_{k-1}$$

for the unknown y_k. It is solved using an LU factorization of A which (together with P) is established once for all k. The algorithm then is:

Inverse Power Method

Input:	A, $n \times n$-matrix, non-singular; $PA = LU$; y_0 n-column; N maximal number of iterations; $\varepsilon > 0$ tolerance
Output:	λ_n eigenvalue, $\lvert\lambda_n\rvert < \lvert\lambda_i\rvert$ for $i \neq n$; x_n eigenvector associated with λ_n

1 Set $\nu_0 := \|y_0\|_\infty$.

2 For $k = 1, 2, \ldots, N$

3 set $y_{k-1} := y_{k-1} \dfrac{1}{\nu_{k-1}}$.

4 Solve $LU y_k := P y_{k-1}$.

5 Determine $\nu_k := \|y_k\|_\infty$.

6 If $\|y_k (\mp) y_{k-1} \nu_k\|_\infty \leq \varepsilon$: go to 7.

7 Set $x_n := y_k \dfrac{1}{\nu_k}$

8 and $\lambda_n := \dfrac{y_k^t y_{k-1}}{y_k^t y_k}$.

Remark 3: For large N one has $A y_N = y_{N-1} \approx y_N \lambda_N$ and thus

$$|\lambda_N| \approx \frac{\|y_{N-1}\|_\infty}{\|y_N\|_\infty} = \frac{1}{\nu_N}\,.$$

Example 1: The table shows the columns $\dfrac{y_k}{\|y_k\|_\infty}$, $k = -3, \ldots, 3$, of the iteration $y_k = A y_{k-1}$ with the matrix

$$A = \begin{bmatrix} 1 & 1 & 0 \\ 4 & -1 & 1 \\ 0 & -1 & 1 \end{bmatrix} = \begin{bmatrix} 1 & & \\ 4 & 1 & \\ 0 & 0.2 & 1 \end{bmatrix} \begin{bmatrix} 1 & 1 & 0 \\ & -5 & 1 \\ & & 0.8 \end{bmatrix} = LU$$

and $y_0 = [0,\ 1,\ 1]^t$:

$$\begin{matrix} -\infty & & -3 & -2 & -1 & k=0 & 1 & 2 & 3 & & k\to\infty & \\ \begin{bmatrix}-0.25\\0\\1\end{bmatrix} & \cdots & \begin{bmatrix}-0.2\\0\\1\end{bmatrix} & \begin{bmatrix}-0.2\\0.2\\1\end{bmatrix} & \begin{bmatrix}0\\0\\1\end{bmatrix} & \begin{bmatrix}0\\1\\1\end{bmatrix} & \begin{bmatrix}1\\0\\0\end{bmatrix} & \begin{bmatrix}0.25\\1\\0\end{bmatrix} & \begin{bmatrix}1\\0\\-0.8\end{bmatrix} & \vdots & \begin{bmatrix}0.33\\1\\-0.33\end{bmatrix} & \begin{bmatrix}1\\0\\-1\end{bmatrix} \end{matrix}.$$

Obviously, the sequence converges to the left side (for $k<0$) to $x_3 = [-0.25,\ 0,\ 1]^t$, but diverges to the right side; there is no dominant eigenvalue. λ_3 is seen to be

$$\frac{1}{\nu_{-3}} = 1 \quad \text{or} \quad \frac{y_{-3}^t y_{-2}}{y_{-3}^t y_{-3}} = 1,$$

respectively where the indices tally with the table.

11.4 Improving an Approximation

The eigenvalues of the matrix $B := A - \lambda_0 I$ are $\mu_i := \lambda_i - \lambda_0$ where λ_i are the eigenvalues of A. Therefore, if an approximation λ_0 for some eigenvalue λ_j of A is known which lies closer to λ_j than to all other λ_i, i.e.,

$$|\lambda_i - \lambda_0| > |\lambda_j - \lambda_0|, \qquad i \neq j,$$

one can determine the eigenvalue $\mu_j := \lambda_j - \lambda_0$ of B by means of the inverse iteration **11.3**, since μ_j is then the eigenvalue with the least modulus of B.

This method which goes back to Wielandt converges fast if λ_0 is close to λ_j. It is also called **fractional iteration.** The map $A \mapsto A - \lambda_0 I$ is known as **spectral shift** because the eigenvalues of A are often referred to as the **spectrum** which is then shifted by $-\lambda_0$.

The eigenvectors are not affected by a spectral shift.

Example 2: The matrix A of Example 1 has the eigenvalue $\lambda_2 = -2$. With the approximation $\lambda_0 = -1$ to λ_2 one gets

$$A - \lambda_0 I = \begin{bmatrix}2 & 1 & 0\\4 & 0 & 1\\0 & -1 & 2\end{bmatrix} = \begin{bmatrix}1 & & \\2 & 1 & \\0 & 0.5 & 1\end{bmatrix} \cdot \begin{bmatrix}2 & 1 & 0\\ & -2 & 1\\ & & 1.5\end{bmatrix}.$$

The table shows the normalized columns of the iteration $y_k = [A - \lambda_0 I]^{-1} y_{k-1}$ for $k = 0, \ldots, 3$

$$\begin{matrix} k=0 & 1 & 2 & 3 & & \infty \\ \begin{bmatrix}0\\1\\0\end{bmatrix} & \begin{bmatrix}-0.5\\1\\0.5\end{bmatrix} & \begin{bmatrix}-0.3\\1.2\\0.3\end{bmatrix} & \begin{bmatrix}-0.35\\1\\0.3\end{bmatrix} & \cdots & \begin{bmatrix}-0.33\\1\\0.33\end{bmatrix} \end{matrix}.$$

Thus, -1 is the eigenvalue of the least modulus of $A - \lambda_0 I$. Hence, A has the eigenvalue $\lambda_2 = \lambda_0 - 1 = -2$. The corresponding eigenvector is $\boldsymbol{x}_2 = [-\frac{1}{3}, 1, \frac{1}{3}]^t$.

11.5 Notes and Exercises

1. If there is no basis of eigenvectors, then, in order to prove Theorem 1, $\boldsymbol{y}_0$ can be decomposed into eigenvectors and generalized eigenvectors.
2. The ratio $|\lambda_1/\lambda_2|$ provides a measurement of the speed at which the method in **11.2** converges.
3. In case of a two-fold eigenvalue $\lambda_1 = \lambda_2$ of maximum modulus, the sequence $\boldsymbol{y}_k$ converges to some eigenvector $\boldsymbol{x}_1 c_1 + \boldsymbol{x}_2 c_2$ associated with $\lambda_1 = \lambda_2$.
4. In case of two different eigenvalues $\lambda_1 = -\lambda_2$ of maximum modulus, the alternate entries of the sequence $\boldsymbol{y}_k$ converge to $\boldsymbol{x}_1 c_1 \pm \boldsymbol{x}_2 c_2$ (cf. Example 1).
5. The moduli of two different eigenvalues of the same modulus can be separated by shifting the spectrum (cf. Example 2).
6. $f(\lambda) := \|A\boldsymbol{x} - \boldsymbol{x}\lambda\|_2$ is minimal at $\lambda = \boldsymbol{x}^t A\boldsymbol{x}/\boldsymbol{x}^t\boldsymbol{x}$, the **Rayleigh quotient**.

12 The LR Algorithm

The classical vector iteration determines merely single eigenvalues with isolated moduli, but together with their eigenvectors. However, there are also methods generating all eigenvalues, but in general without the eigenvectors.

12.1 The Algorithm of Rutishauser

Rutishauser had the idea in 1958 to multiply the factors of a LU decomposition of a non-singular matrix A in reverse order and to repeat this action:

$$L_1 U_1 := A_1 := A \quad , \quad A_2 := U_1 L_1 \, ,$$

$$\vdots$$

$$L_k U_k := A_k \quad , \quad A_{k+1} := U_k L_k \, .$$

$$\vdots$$

If this is feasible, i.e., if each matrix $A_k = U_{k-1} L_{k-1}$ admits an LU factorization, then the following turns out to be true:

(1) The matrices A_k are similar to A.

(2) If the eigenvalues of A have pairwise distinct moduli then the matrices L_k converge to the identity matrix I and the matrices U_k towards an upper triangular matrix U.

Hence, the diagonal of U contains the eigenvalues of A. In general, they are even ordered with respect to their moduli.

This method leads to the following algorithm:

LR Algorithm

Input:	A, $n \times n$-matrix, non-singular; N; $\varepsilon > 0$ tolerance
Output:	$A := U$ triangular matrix similar to A

For $k = 1, 2, \ldots, N$
  Decompose $A = LU$,
  if a factorization is impossible: failure.
  If $\|L - I\|_\infty \leq \varepsilon \|A\|_\infty$: **stop.**
  Form $A = UL$

The algorithm fails if some eigenvalues have the same modulus and also if one of the LU factorizations does not exist. Pivoting would destroy the similarity and is not permitted.

12.2 Proving Convergence

First (1) is proved. A simple manipulation yields

$$A_2 = U_1 L_1 = L_1^{-1} L_1 U_1 L_1 = L_1^{-1} A_1 L_1 ,$$

and analogously

$$A_{k+1} = [L_1 \cdots L_k]^{-1} A_1 [L_1 \cdots L_k] ,$$

which verifies property (1).

The proof of (2) most often given today goes back to Wilkinson. It is based on the comparison of two different representations of the unique LU factorization of A^k:

1. The LU factorization can be expressed in terms of the L_k and U_k of the LR algorithm:

$$
\begin{aligned}
A^k &= (L_1U_1)^k = L_1(U_1L_1)^{k-1}U_1 = L_1A_2^{k-1}U_1 \\
&= L_1L_2(U_2L_2)^{k-2}U_2L_1 = L_1L_2A_3^{k-2}U_2U_1 \\
&\ \vdots \\
&= [L_1\cdots L_k][U_k\cdots U_1]\,.
\end{aligned} \tag{3}
$$

2. Another representation of the same factorization is obtained after a somewhat longer calculation. Because the eigenvalues λ_i of A have different moduli the modal matrix X exists, i.e., with the notation of **11**,

$$X^{-1}AX = \Lambda\,, \qquad |\lambda_1| > \cdots > |\lambda_n| > 0\,.$$

To simplify the proof it is assumed that the LU factorization of X and $Y = X^{-1}$ exist. Let

$$X = L_xU_x\,, \qquad Y = L_yU_y\,.$$

This gives

$$A^k = X\Lambda^kY = L_xU_x\Lambda^kL_y\Lambda^{-k}\Lambda^kU_y\,.$$

The product $\Lambda^kL_y\Lambda^{-k}$ is a lower triangular matrix which arises from L_y by multiplying each entry i,j of L_y for $i>j$ by the factor $(\lambda_i/\lambda_j)^k$. Because of $|\lambda_1|>\cdots>|\lambda_n|$, it follows $|\lambda_i/\lambda_j|<1$ and

$$\Lambda^kL_y\Lambda^{-k} = I + \Delta_k \text{ with } \lim_{k\to\infty}\Delta_k = 0\,.$$

This establishes the identities

$$
\begin{aligned}
A^k &= L_xU_x[I+\Delta_k]\Lambda^kU_y \\
&= L_x[I + U_x\Delta_kU_x^{-1}]U_x\Lambda^kU_y\,.
\end{aligned}
$$

The matrices $I + U_x\Delta_kU_x^{-1}$ converge to the identity matrix as well. Therefore, these matrices admit an LU factorization $L_{(k)}U_{(k)}$ for sufficiently large k with

$$\lim_{k\to\infty} L_{(k)} = I \qquad \text{and} \qquad \lim_{k\to\infty} U_{(k)} = I\,.$$

So it ensues

$$A^k = L_xL_{(k)}U_{(k)}U_x\Lambda^kU_y\,. \tag{4}$$

3. Comparing (3) and (4) shows because of the uniqueness of the LU factorization of A^k and A^{k-1}, that

$$L_k = [L_1\cdots L_{k-1}]^{-1}[L_1\cdots L_k] = [L_xL_{(k-1)}]^{-1}[L_xL_{(k)}] = L_{(k-1)}^{-1}L_{(k)}\,.$$

Thus, L_k converges to the identity matrix. Analogously one verifies that

$$\begin{aligned} U_k &= [U_k \cdots U_1][U_{k-1} \cdots U_1]^{-1} \\ &= [U_{(k)} U_x \Lambda^k U_y][U_{(k-1)} U_x \Lambda^{k-1} U_y]^{-1} \\ &= U_{(k)} U_x \Lambda U_x^{-1} U_{(k-1)}^{-1}, \end{aligned}$$

i.e., U_k converges to $U_x \Lambda U_x^{-1}$. The product of two upper triangular $n \times n$-matrices $U = [u_{i,j}]$ and $V = [v_{i,j}]$ is an upper triangular matrix $W = [w_{i,j}] = UV$. Moreover, $w_{i,i} = u_{i,i}\, v_{i,i}$ for all i. Thus, corresponding diagonal entries of U_x and U_x^{-1} multiply to 1 and $U_x \Lambda U_x^{-1}$ is an upper triangular matrix with the eigenvalues $\lambda_1, \ldots, \lambda_n$ in the diagonal. The eigenvalues are even ordered with respect to their moduli. However, this property is lost if X and X^{-1} do not have an LU factorization.

12.3 Pairs of Eigenvalues with Equal Modulus

The algorithm fails in case two eigenvalues have the same modulus, e.g., $|\lambda_{j+1}| = |\lambda_j|$. The subdiagonal element $d_{j+1,j}$ of the matrix

$$D_k := \Lambda^k L_y \Lambda^{-k} = I + \Delta_k$$

does not become arbitrarily small then. Taking this into account, one comes after only slightly more extensive considerations to the following result.

Even for large k the entry $l_k := l_{j+1,j}$ of L_k on the subdiagonal is substantially distinct from zero

$$L_k = \begin{bmatrix} 1 & & & & \\ & \ddots & & & \\ & & 1 & & \\ & & l_k & 1 & \\ & & & & \ddots & \\ & & & & & 1 \end{bmatrix}.$$

For this reason, the entry $q_k := a_{j+1,j}$ on the subdiagonal of A_k for large k, does not vanish, either

$$A_k = L_k U_k = \begin{bmatrix} * & \cdots & * & * & \cdots & * \\ & \ddots & \vdots & \vdots & & \vdots \\ & & * & & & \\ & & q_k & * & & \\ & & & & \ddots & \vdots \\ & & & & & * \end{bmatrix}$$

and the columns j and $j+1$ of A_k do not converge. But the similarity of A_k and A ensures that the diagonal entries $a_{i,i}$, $i \neq j, j+1$, and the eigenvalues of the submatrix

$$\begin{bmatrix} a_{j,j} & a_{j,j+1} \\ a_{j+1,j} & a_{j+1,j+1} \end{bmatrix}$$

converge to the eigenvalues of A.

Example 1: Consider the Example 1 of **11.3**. It is

$$A_1 = \begin{bmatrix} 1 & 1 & 0 \\ 4 & -1 & 1 \\ 0 & -1 & 1 \end{bmatrix} = \begin{bmatrix} 1 & & \\ 4 & 1 & \\ 0 & 0.2 & 1 \end{bmatrix} \begin{bmatrix} 1 & 1 & 0 \\ & -5 & 1 \\ & & 0.8 \end{bmatrix} = L_1 U_1 .$$

Calculating with 2 digits after the decimal point one obtains after four iterations

$$A_5 = \begin{bmatrix} 1 & 1 & 0 \\ 3.04 & -1 & 1 \\ 0 & -0.05 & 1 \end{bmatrix} .$$

The first two eigenvalues can be computed from the quadratic equation

$$\det \begin{bmatrix} 1-\lambda & 1 \\ 3.04 & -1-\lambda \end{bmatrix} = 0 .$$

They are $\lambda_1 = +2.01$ and $\lambda_2 = -2.01$. The approximation $\lambda_3 = 1$ can be read off directly from A_5. The exact values are $\lambda_1 = 2$, $\lambda_2 = -2$, $\lambda_3 = 1$.

Remark 1: The algorithm converges rather slowly and is very expensive. Therefore it should only be used with tridiagonal matrices. This is done, e.g., in **17** for the QD method.

12.4 Notes and Exercises

1. In case of three or more eigenvalues with equal moduli three columns of A_k or more, respectively, do not converge. However, the eigenvalues appear on the diagonal or are the eigenvalues of 3×3 or larger principal submatrices.
2. Analogous to the LR algorithm one can construct a **QU algorithm** (Francis 1961): One forms
$$A_k = Q_k U_k , \qquad A_{k+1} := U_k Q_k .$$
The QU factorization of A_k is always possible if A is non-singular.
3. Only the diagonal of the A_k converges in the QU algorithm provided the moduli of all eigenvalues are pairwise distinct. It converges to the (ordered) eigenvalues of A.
4. If two (and more) eigenvalues have the same modulus then the eigenvalues of respective submatrices of A_k converge to these eigenvalues in the QU algorithm.

13 One-Dimensional Iteration

The iterative solution of a problem can be interpreted as the determination of a fixed point of some mapping. The related convergence behavior will be studied for the class of **contractive mappings** which allow for a general analysis.

13.1 Contractive Mappings

The iteration rule for one variable x can be written as

$$x_{k+1} = f(x_k).$$

Depending on the function f and the starting value x_0 the sequence x_k, thus defined, may or may not converge to a **fixed point** $s = f(s)$.

A mapping $f : x \mapsto f(x)$ is called **contractive** on the closed interval $I := [a, b]$ if it satisfies the following two conditions:

1. The image of the interval I lies in I, i.e.,

$$f(I) \subset I.$$

2. There exists a **Lipschitz constant** $0 \le L < 1$ such that for all $x, y \in I$

$$|f(x) - f(y)| \le L|x - y|.$$

This implies that f is continuous and leads to the **Fixed-point Theorem:**

Theorem: *A mapping f which is contractive on the interval $I = [a, b]$ has exactly one fixed point $s = f(s)$ in I. It is the limit of every sequence $x_{k+1} = f(x_k)$ formed with an arbitrary starting value $x_0 \in I$.*

The proof is carried out in three steps:

Convergence: The distance $|x_{k+1} - x_k|$ can be estimated by the second condition

$$|x_{k+1} - x_k| \le L|x_k - x_{k-1}| \le \cdots \le L^k|x_1 - x_0|.$$

Therefore, one gets for $n > k$

$$\begin{aligned} |x_n - x_k| &\le |x_n - x_{n-1}| + \cdots + |x_{k+1} - x_k| \\ &\le (L^{n-1} + \cdots + L^k)|x_1 - x_0| \\ &= L^k(L^{n-1-k} + \cdots + 1)|x_1 - x_0|. \end{aligned}$$

The formula $1 + L + L^2 + \cdots = \dfrac{1}{1-L}$ for a geometric series then establishes

$$(1) \qquad |x_n - x_k| \le \frac{L^k}{1-L}|x_1 - x_0|.$$

Thus, for every $\varepsilon > 0$ there is a k such that, for all $n > k$, $|x_n - x_k| < \varepsilon$, i.e., the x_k form a (convergent) Cauchy sequence. The limit s lies in I because each x_k lies in I by condition 1 and because I is closed.

Existence: Since f is continuous, one has

$$f(s) = f\left(\lim_{k\to\infty} f^k(x_0)\right) = \lim_{k\to\infty} f^{k+1}(x_0) = s,$$

i.e., s is a fixed point.

Uniqueness: Two fixed points $r = f(r)$ and $s = f(s)$ satisfy

$$|r - s| = |f(r) - f(s)| \le L|r - s|.$$

Since $L < 1$, $r = s$ follows. This completes the proof.

Remark 1: The iteration rule $x_{k+1} = f(x_k)$ has an appealing geometric interpretation:

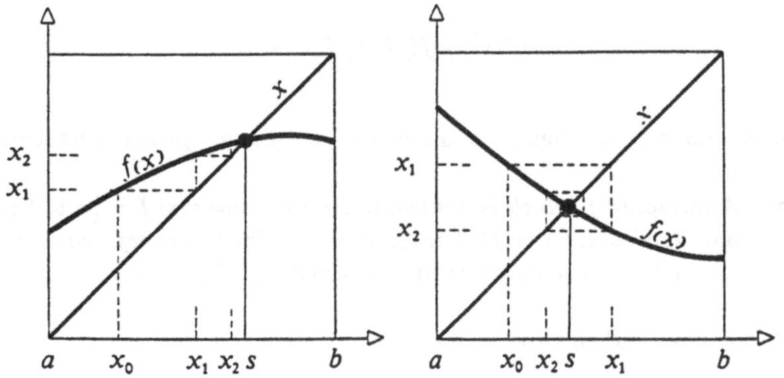

Figure 13.1 Geometric interpretation

Example 1: The function $f(x) = \cos x$ is contractive in $I = [-1, 1]$; a suitable Lipschitz constant is $L = \sin 1 < 1$.

13.2 Error Bounds

The distance between $x_k = f(x_{k-1})$ and s can be estimated in two ways provided f is contractive:

The a priori bound

$$|x_k - s| \leq \frac{L^k}{1-L}|x_1 - x_0|$$

follows immediately from (1) for $n \to \infty$ and leads to the

a posteriori bound

$$|x_k - s| \leq \frac{L}{1-L}|x_k - x_{k-1}|$$

if x_{k-1} is interpreted as a starting value for another iteration. By means of the a priori bound one can, at the beginning of the iteration, calculate an upper bound

$$\frac{\log \varepsilon(1-L) - \log|x_1 - x_0|}{\log L}$$

for the number N of iterations necessary to achieve a certain accuracy. Generally, this bound is much too large. The a posteriori bound is a bound of how far x_k is away from s. It can be used to measure the accuracy at each step of the iteration:

Iteration in One Variable

Input: $f(x)$ contractive in I with $L < 1$; $x_0 \in I$; $\varepsilon > 0$; N
Output: $s = f(s)$ fixed point

1 For $k = 1, 2, \ldots, N$
2 determine $x_k := f(x_{k-1})$.
3 If $|x_k - x_{k-1}| \leq \varepsilon \frac{1-L}{L}$: go to 4.
4 Set $s := x_k$.

Remark 2: Often L itself is found through an estimation. It should be chosen as small as possible.

13.3 Rate of Convergence

The number of iterations needed to determine the limit s within a set tolerance is linked to the **rate of convergence**, that is how much the accuracy is improved in every step.

In order to get more insight, consider the Taylor series of f about s:

$$f(x) = f(s) + f'(s)(x-s) + \frac{1}{2}f''(s)(x-s)^2 + \cdots,$$

where it is assumed that f is sufficiently often differentiable.

Introducing the abbreviations $d_k := x_k - s$ for the error and $c_i := \frac{1}{i!}f^{(i)}(s)$ for the coefficients gives

$$d_{k+1} = c_1 d_k + c_2 d_k^2 + \cdots = (c_1 + c_2 d_k + \cdots) d_k .$$

In case $c_1 = f'(s) \neq 0$ the error d_{k+1} depends linearly on d_k for large k and small d_k. It can converge to zero only if $|c_1| < 1$. Then, this is termed **linear convergence** with the **convergence factor** c_1.

In case $c_1 = 0$, but $c_2 = \frac{1}{2}f''(s) \neq 0$ one speaks of **quadratic convergence.**

In general, a sequence x_k is said to be **convergent of order** p if p is the minimal positive constant with

$$\lim_{k\to\infty} \frac{d_{k+1}}{d_k^p} = c_p \neq 0 .$$

Example 2: The sequence $x_{k+1} = \cos x_k$ converges linearly in $I = [-1, 1]$.

Example 3: The sequence $x_{k+1} = x_k(2 - x_k)$ converges quadratically to $s = 1$ in $I = [\frac{1}{3}, \frac{4}{3}]$.

13.4 Aitken's Δ^2-Method

In case of linear convergence and for large k there are the approximations:

$$x_{k+1} - s \approx c_1(x_k - s), \text{ and}$$
$$x_{k+2} - s \approx c_1(x_{k+1} - s).$$

Eliminating c_1 in these two "equations" yields a new approximation x_k^* for s, namely

$$s \approx x_k^* := x_k - \frac{(x_{k+1} - x_k)^2}{x_{k+2} - 2x_{k+1} + x_k}.$$

Using the common abbreviations

$$\Delta x_k := x_{k+1} - x_k \text{ and } \Delta^2 x_k := \Delta(\Delta x_k)$$

for the **forward differences**, this can be written as

$$x_k^* := x_k - \frac{\Delta x_k}{\Delta^2 x_k}\Delta x_k .$$

This improvement of x_k was introduced by Aitken, 1926. Under certain assumptions, the sequence $x_0^*, x_1^*, x_2^*, \ldots$ converges more rapidly to s than the original sequence $x_0, x_1, x_2, \ldots$. A further discussion is omitted here. However, it should be mentioned that the method is not restricted to sequences which stem from an iteration.

It seems reasonable to take x_0^* as a new starting value as soon as x_1 and x_2 are computed. This gives the algorithm of Steffensen (1933):

Steffensen's Iteration

Input: $f(x)$ contracting in I; $x_0 \in I$; $\eta > 0$; N
Output: $s = f(s)$ fixed point

1 For $k = 1, 2, \ldots, N$
2 determine $x_1 := f(x_0)$
3 and $x_2 := f(x_1)$.
4 Set $x_0 := x_0 - \dfrac{\Delta x_0}{\Delta^2 x_0} \Delta x_0$.
5 If $\left| \dfrac{\Delta x_0}{\Delta^2 x_0} \Delta x_0 \right| \leq \eta$: go to **6** .
6 Set $s := x_0$.

Example 4: For $f(x) = \cos x$, $x_0 = 0.750$ and $\eta = 0.001$ Steffensen's iteration terminates after two steps:

$$\begin{array}{lcl} x_0 = 0.750 & \rightarrow 0.740 & \rightarrow 0.739 = s \\ x_1 = 0.732 & 0.738 & \\ x_2 = 0.744 & 0.740 & \end{array} .$$

13.5 Geometric Acceleration

A fixed point s of $f(x)$ is also a fixed point of the mapping

$$\bar{f}(x) := x - \mu(x)(x - f(x)), \quad \mu(x) \neq 0 \text{ in } I,$$

and vice versa. But $\bar{f}$ is not necessarily contractive in I if f is.

The convergence factor $\bar{c}_1$ of $\bar{f}$ depends on the convergence factor c_1 of f, namely

$$\bar{c}_1 := \bar{f}'(s) = 1 - \mu(s)(1 - f'(s)) - \mu'(s)(s - f(s)) = 1 - \mu(s)(1 - f'(s)) .$$

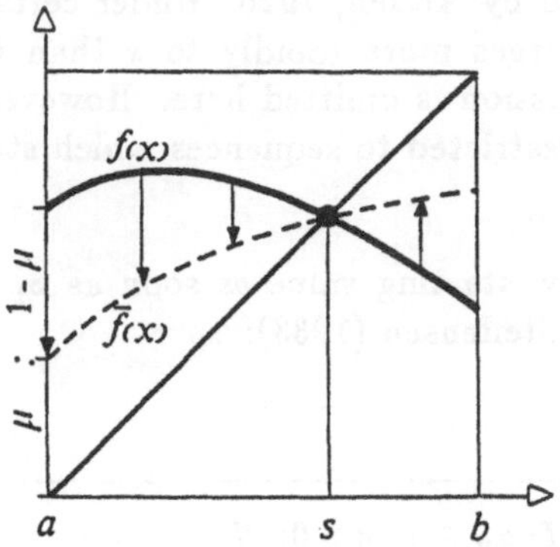

Figure 13.2
Geometric acceleration

$\bar{c}_1$ vanishes for $\mu(s) = 1/(1-c_1)$ and the sequence $\bar{x}_{k+1} = \bar{f}(x_k)$ converges, if at all, at least quadratically. There are several choices for $\mu(x) := 1/(1-c(x))$ which satisfy this condition for s or almost satisfy it. They result in different methods which converge quadratically or almost quadratically.

Whittaker's method: A particular simple but nevertheless quick method is obtained by setting $c =$ constant and approximately $c \approx \pm L$. This method converges quadratically only if $c = c_1$.

Steffensen's method: The iteration scheme in **13.4** is obtained by setting

$$c(x) = \frac{f(f(x)) - f(x)}{f(x) - x}.$$

It converges quadratically.

Example 5: For $f(x) = \cos x$ and $c = -\frac{1}{2} =$ constant, one gets $\mu = \frac{2}{3}$ and, hence,

$$\bar{f} = \frac{1}{3}x + \frac{2}{3}\cos x.$$

With $\bar{x}_0 = 0.75$ one gets $\bar{x}_1 = 0.738$ and already $\bar{x}_2 = 0.739$.

13.6 Roots

One of the most important applications of convergence acceleration is that of finding a zero of a function $g(x)$ as the fixed point of the mapping

$$f(x) := x - \mu(x)g(x).$$

The convergence factor is $c_1 := f'(s) = 1 - \mu(s)g'(s)$. The fixed-point iteration converges, if at all, quadratically with $g'(s) \neq 0$ and $\mu(s) = 1/g'(s)$. Different $\mu(x)$ result in different methods:

Whittaker's method: One sets

$$\mu(x) = m = \text{constant}\,.$$

Regula falsi: This well-known method is obtained for

$$\mu(x_k) = \frac{\Delta x_{k-1}}{\Delta g_{k-1}}$$

where $g_{k-1} = g(x_{k-1})$. It takes two starting values.

Newton-Raphson method: This classical method is obtained for

$$\mu(x) = \frac{1}{g'(x)}\,.$$

In general, it converges quadratically.

The corresponding algorithm with a suitable subroutine evaluating μ at x_k has the form:

Root-Finding

Input:	$g(x)$; $\mu(x)$; x_0; η; N
Output:	s with $g(s) = 0$ root

1 For $k = 1, 2, \ldots, N$

  2 determine $g := g(x_0)$

  3 and $\mu := \mu(x_0)$.

4 Set $x_0 := x_0 - \mu g$.

5 If $|\mu g| \leq \eta$: go to 6 .

6 Set $s := x_0$.

Remark 3: There is a variety of sufficient conditions ensuring convergence for these methods, especially for the Newton-Raphson method. However, in non-trivial applications these conditions can be checked only with difficulties. The essence of these conditions is always:

If $g'(s) \neq 0$, then the root has a neighborhood such that these methods converge to s for all starting values x_0 from this neighborhood.

Remark 4: $1/\mu(x_k)$ is the slope of the line connecting the point (x_k, g_k) on g with the point $(x_{k+1}, 0)$ on the x-axis.

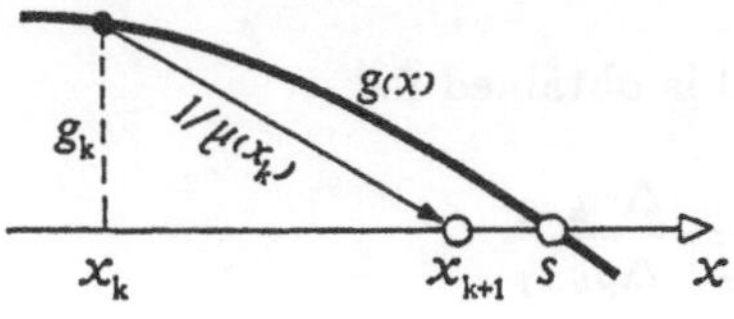

Figure 13.3
Root finding

Example 6: $g(x) = x - \cos x$ gives $g'(x) = 1 + \sin x$. For $x_0 = 0.75$ one gets already after one step of the Newton-Raphson iteration

$$x_1 = 0.75 - \frac{0.018}{1.68} = 0.739\,.$$

13.7 Notes and Exercises

1. The maximum slope of a chord of f in I is a Lipschitz constant L for f.
2. $L \geq \max_{x \in I} |f'(x)|$ for every Lipschitz constant L.
3. Aitken's Δ^2-process has a geometric interpretation: It replaces $f(x)$ between x_{k+1} and x_k by a chord.
4. The accuracy of the approximation is improved by approximately one decimal digit in every step for $c_1 = 0.1$.
5. The accuracy is roughly doubled in each step if $c_1 = 0$ and $c_2 = 1$.
6. The convergence of a sequence produced by the Δ^2-process can again be accelerated using the Δ^2-process once more.
7. $\Delta^2 x_k = 0$ implies $\Delta x_{k+1} = \Delta x_k$ and therefore $L \geq 1$.
8. Use Steffenson's method to find a root. This procedure is related to the regula falsi.
9. The Newton-Raphson method fails if $g'(s) = 0$. However if $g''(s) \neq 0$ one can use

$$\mu(x) := \frac{2}{g''(x)}\,.$$

14 Multi-Dimensional Iteration

The reasoning for the one-dimensional iteration applies straightforwardly to more dimensional iteration as well. The scalar x is replaced by an n-column $\boldsymbol{x}$, the absolute value by some norm, etc.

14.1 Contractive Mappings

A mapping $\boldsymbol{f}: \mathbb{R}^n \to \mathbb{R}^n$ with the components $f_i : \mathbb{R}^n \to \mathbb{R}^n$ is called **contractive** in the cube $\boldsymbol{I} := [\boldsymbol{a}, \boldsymbol{b}]$ if for some norm the following two conditions hold:

1. $\boldsymbol{f}(\boldsymbol{I}) \subset \boldsymbol{I}$,
2. $||\boldsymbol{f}(\boldsymbol{x}) - \boldsymbol{f}(\boldsymbol{y})|| \leq L||\boldsymbol{x} - \boldsymbol{y}||$, for all $\boldsymbol{x}, \boldsymbol{y} \in \boldsymbol{I}$ and $0 \leq L < 1$, L fixed.

Further, there is also a multi-dimensional **Fixed-point Theorem.**

Theorem: *A mapping $\boldsymbol{f}$ which is contractive in $\boldsymbol{I}$ has exactly one fixed point $\boldsymbol{s}$ in $\boldsymbol{I}$. This fixed point is the limit of the sequence $\boldsymbol{x}_k = \boldsymbol{f}(\boldsymbol{x}_{k-1})$ for every starting value $\boldsymbol{x}_0 \in \boldsymbol{I}$. This distance $||\boldsymbol{x}_k - \boldsymbol{s}||$ can be estimated by the* **a priori bound**

$$||\boldsymbol{x}_k - \boldsymbol{s}|| \leq \frac{L^k}{1-L}||\boldsymbol{x}_1 - \boldsymbol{x}_0||$$

and by the **a posteriori bound**

$$||\boldsymbol{x}_k - \boldsymbol{s}|| \leq \frac{L}{1-L}||\boldsymbol{x}_k - \boldsymbol{x}_{k-1}||.$$

The proofs can be copied word for word from the one-dimensional case in **13**. The ε-neighborhood will depend on the norm, though. Also, the algorithm there can be used here as well.

14.2 Rate of Convergence

Here, too, general statements about the rate of convergence can be derived from the Taylor expansion of $\boldsymbol{f}$ about $\boldsymbol{s}$. Assuming sufficient differentiability it is

$$\boldsymbol{f}(\boldsymbol{x}) = \boldsymbol{f}(\boldsymbol{s}) + D_{\boldsymbol{f}}(\boldsymbol{s})(\boldsymbol{x} - \boldsymbol{s}) + \cdots,$$

where $D_{\boldsymbol{f}} := \left[\frac{\partial f_i}{\partial \boldsymbol{x}_k}\right]$ denotes the **Jacobian matrix** of $\boldsymbol{f}$. Using the abbreviation $\boldsymbol{d}_k := \boldsymbol{x}_k - \boldsymbol{s}$ for the error, one gets

$$\boldsymbol{d}_{k+1} = D_{\boldsymbol{f}}(\boldsymbol{s})\boldsymbol{d}_k + \cdots.$$

As in **13** one speaks of **linear convergence** if $D_f(s) \neq O$ and of at least **quadratic convergence** otherwise. $D_f(s)$ is called the **iteration matrix.** A sufficient condition for convergence is that $\|D_f(s)\| < 1$.

More general, a sequence x_k is said to be **convergent of order p** if p is the least positive constant such that

$$\lim_{k\to\infty} \frac{\|d_{k+1}\|}{\|d_k\|^p} = a_p \neq 0 .$$

The order of convergence does not depend on the norm.

Example 1: The classical vector iteration in **11** converges linearly.

Example 2: The relaxation methods in **8** converge linearly.

14.3 Accelerating the Convergence

A fixed point s of f is also a fixed point of the mapping

$$\bar{f}(x) := x - M(x)[x - f(x)], \quad M(x)\ n \times n\text{-matrix, non-singular in } I,$$

and vice versa. However, $\bar{f}$ is not necessarily contractive in I if f is.

The iteration matrix $D_{\bar{f}}(s)$ of $\bar{f}$ is obtained through differentiation

$$D_{\bar{f}}(s) = I - M(s)[I - D_f(s)] .$$

If $M(s) = [I - D_f(s)]^{-1}$, then $D_{\bar{f}}(s)$ vanishes and the sequence $\bar{x}_k = \bar{f}(\bar{x}_{k-1})$ converges, if it converges, at least quadratically. For different choices of $M(x) = [I - D(x)]^{-1}$, which satisfy this condition at s or almost satisfy it, one obtains again different methods that converge quadratically or almost quadratically:

Whittaker's method: One sets $D(x_k) = D =$ constant, especially $D \approx D_f(s)$.

Steffensen's method: One sets $D(x_0) := \Delta X_1[\Delta X_0]^{-1}$ with the notation $\Delta X_1 := X_2 - X_1$ and $X_1 := [x_1, \ldots, x_n]$, the $n \times n$-matrix of the $x_1 = f(x_0), \ldots,$ $x_n = f(x_{n-1})$ etc.

Remark 1: The iteration $x_{k+1} = \bar{f}(x_k)$ is better written as a linear system

$$[I - D(x_k)][x_{k+1} - x_k] = f(x_k) - x_k$$

for the difference $x_{k+1} - x_k$.

14.4 Roots of Systems

The acceleration of the convergence has an important application in finding the root $\boldsymbol{s}$ of a system $\boldsymbol{g}(\boldsymbol{x}) = \boldsymbol{o}$ of n equations $g_i(\boldsymbol{x}) = 0$ as the fixed point of the mapping

$$\boldsymbol{f}(\boldsymbol{x}) := \boldsymbol{x} - M(\boldsymbol{x})\boldsymbol{g}(\boldsymbol{x}),$$

where $M(\boldsymbol{x})$ is an $n \times n$-matrix, non-singular in $\boldsymbol{I}$. There are several iteration methods arising from different $M(\boldsymbol{x})$. They converge at least quadratically if $M(\boldsymbol{s}) = [D_g(\boldsymbol{s})]^{-1}$. In particular there are:

Whittaker's method using $M(\boldsymbol{x}) = M =$ constant, especially $M \approx [D_g(\boldsymbol{s})]^{-1}$.

Newton method using $M(\boldsymbol{x}) := [D_g(\boldsymbol{x})]^{-1}$.

Remark 2: The Newton method, too, is better written as a linear system

$$D_g(\boldsymbol{x}_k) \cdot \Delta\boldsymbol{x}_k + \boldsymbol{g}(\boldsymbol{x}_k) = \boldsymbol{o}$$

for the difference $\Delta\boldsymbol{x}_k := \boldsymbol{x}_{k+1} - \boldsymbol{x}_k$. The inversion is unnecessary, then.

Remark 3: Remark 3 in **13.6** about sufficient conditions for the convergence applies analogously to systems as well.

Example 3: A two-dimensional Newton iteration is used for the method of Bairstow in **15.5**.

14.5 Notes and Exercises

1. Newton's method replaces the surfaces $z = f_i(\boldsymbol{x})$ by the tangent planes at $\boldsymbol{x}_k$ and thus the intersection of these surfaces by the intersection of those tangent planes.
2. The Newton method solves every linear system $A\boldsymbol{x} - \boldsymbol{a} = \boldsymbol{o}$ in one step. The Jacobian matrix of the system is A itself.
3. In order to expand the region of convergence of Newton's method one often uses the modification
$$\boldsymbol{x}_{k+1} = \boldsymbol{x}_k - \lambda[D_g(\boldsymbol{x}_k)]^{-1}\boldsymbol{g}(\boldsymbol{x}_k)$$
where $\lambda \in (0, 1]$ is determined such that $\|\boldsymbol{g}(\boldsymbol{x}_{k+1})\| < \|\boldsymbol{g}(\boldsymbol{x}_k)\|$. With this one comes close to the root in a few steps. Thence the root is determined with $\lambda = 1$.

15 Roots of Polynomials

An expression of the form

(1) $$\mathrm{pol}(\lambda) := a_0\lambda^n + a_1\lambda^{n-1} + \cdots + a_{n-1}\lambda + a_n\,, \quad a_0 \neq 0\,,$$

is referred to as a **polynomial** of degree n in λ. In particular, if $a_0 = 1$ it is called **monic**. Because of their simple form, polynomials are easy to deal with.

15.1 The Horner Scheme

A polynomial of degree n in λ can be written as

$$\mathrm{pol}(\lambda) = (\cdots((a_0\lambda + a_1)\lambda + a_2)\lambda + \cdots + a_{n-1})\lambda + a_n$$

and be evaluated for any fixed $\lambda = \lambda_0$ in this manner. One determines successively the values p_k of the parentheses going from the inside to the outside. This results in the following algorithm:

Horner

Input: $a_0, a_1, \ldots, a_n;\ \lambda_0$

Output: $p := a_0\lambda_0^n + a_1\lambda_0^{n-1} + \cdots + a_{n-1}\lambda_0 + a_n$

Set $p := a_0$.

For $k = 1, 2, \ldots, n$

  set $p := p\lambda_0 + a_k$.

The manual computation is facilitated by the well-known **Horner scheme**. The a_k are written in a row. The values p_k are placed underneath during the calculations:

λ_0	a_0	a_1	$\ldots$	a_n
0	p_0	p_1	$\ldots$	$\underline{p_n = p}$

The rule for the computation of the $p_k = p_{k-1}\lambda_0 + a_k$ can be shown graphically as follows:

(2)

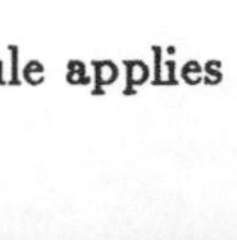

If one defines $p_{-1} := 0$, then this rule applies also to $p_0 = a_0$.

15.2 The Extended Horner Scheme

The polynomial (1) has the derivative

$$\mathrm{pol}'(\lambda) = na_0\lambda^{n-1} + (n-1)a_1\lambda^{n-2} + \cdots + a_{n-1}\,.$$

It could be evaluated for $\lambda = \lambda_0$ as above. However, there is a simpler way. Putting down the terms $p_0, \ldots, p_{n-1}$ with respect to the a_k and multiplying them by $\lambda_0^{n-1}, \ldots, \lambda_0, 1$

$$\begin{array}{ll|l} p_0 & = a_0 & \cdot\lambda_0^{n-1} \\ \vdots & \quad \vdots & \vdots \\ p_{n-1} & = a_0\lambda_0^{n-1} + \cdots + a_{n-1} & \cdot 1, \end{array}$$

one can observe that they sum to $\mathrm{pol}'(\lambda_0)$, i.e.,

$$\mathrm{pol}'(\lambda_0) = p_0\lambda_0^{n-1} + p_1\lambda_0^{n-2} + \cdots + p_{n-1}\,.$$

The p_k are polynomials in λ_0. Their values appear in the second row of the Horner scheme. Writing underneath the

$$q_k := q_{k-1}\lambda_0 + p_k$$

gives the **extended Horner scheme:**

λ_0	a_0	a_1	$\ldots$	a_{n-1}	a_n
0	p_0	p_1	$\ldots$	p_{n-1}	$\underline{p_n = p}$
0	q_0	q_1	$\ldots$	$\underline{q_{n-1} = q}$	

in which $q_{n-1} = q = \mathrm{pol}'(\lambda_0)$. The second row is determined by the same rule (2). Here, too, the rule yields $q_0 = p_0$ if one sets $q_{-1} := 0$. Altogether, the algorithm for computing $\mathrm{pol}(\lambda_0)$ and $\mathrm{pol}'(\lambda_0)$ is:

Horner, Extended

Input: $a_0, a_1, \ldots, a_n;\ \lambda_0$

Output: $p := a_0\lambda_0^n + \cdots + a_n;\ q := na_0\lambda_0^{n-1} + \cdots + a_{n-1}$

```
Set q := 0, p := a_0 .
For k = 1, 2, ..., n
    set  q := q λ_0 + p
    and p := p λ_0 + a_k .
```

15.3 Simple Roots

Simple zeros of a polynomial pol(λ) for which a suitable approximation λ_0 is known can be determined iteratively by Newton's method in **13.6**. Namely, one forms the sequence

$$\lambda_{k+1} := \lambda_k - \frac{\mathrm{pol}(\lambda_k)}{\mathrm{pol}'(\lambda_k)}$$

where pol(λ_k) and pol$'(\lambda_k)$ are determined in the extended Horner scheme:

Simple Root

Input: $a_0, a_1, \ldots, a_n, \lambda_0;\ \varepsilon > 0;\ N$

Output: λ with $a_0\lambda^n + \cdots + a_n \approx 0$

```
Set λ = λ0 .
For i = 1, 2, ..., N
    determine p(λ), q(λ) by [Horner, Extended.]
    If |p/q| ≤ ε: stop.
    Set λ := λ − p/q .
```

Example 1: The approximation $\lambda_0 = 0.95$ of the zero $\lambda = 1$ of the polynomial

$$\mathrm{pol}(\lambda) = \lambda^3 - \lambda^2 - 4\lambda + 4$$

is to be improved by one Newton iteration. The extended Horner scheme is

0.95	1	−1	−4	4
0	1	−0.05	−4.05	0.15 = p
0	1	0.95	−3.19 = q	

and therefore the improved approximation is

$$\lambda_1 = 0.95 + \frac{0.15}{3.19} = 0.997\,.$$

15.4 Bairstow's Method

If the real polynomial (1) has a complex root μ, then it is known that the corresponding conjugate complex number $\bar{\mu}$ is a root of the polynomial as well. Both numbers are the roots of a real quadratic polynomial $\lambda^2 - u\lambda - v$. Using the Euclidean algorithm, one can divide the original polynomial by $\lambda^2 - u\lambda - v$ without remainder. Now, in general, with an arbitrary factor $\lambda^2 - u\lambda - v$ one gets

(3) $\mathrm{pol}(\lambda) = (b_0\lambda^{n-2} + b_1\lambda^{n-3} + \cdots + b_{n-2})(\lambda^2 - u\lambda - v) + b_{n-1}(\lambda - u) + b_n\,.$

The coefficients b_{n-1}, b_n depend on u and v. If $b_{n-1}(u,v) = b_n(u,v) = 0$ then $\lambda^2 - u\lambda - v$ is a factor of the polynomial (3) and the roots of $\lambda^2 - u\lambda - v$ are roots of (3).

Bairstow's method is based on the idea to improve an approximation u_0, v_0 of the common root u, v of b_{n-1} and b_n using Newton's method.

The notation

$$\boldsymbol{x} := \begin{bmatrix} u \\ v \end{bmatrix}, \quad \boldsymbol{g} := \begin{bmatrix} b_n \\ b_{n-1} \end{bmatrix}, \quad D_{\boldsymbol{g}} := \begin{bmatrix} \dfrac{\partial b_n}{\partial u} & \dfrac{\partial b_n}{\partial v} \\ \dfrac{\partial b_{n-1}}{\partial u} & \dfrac{\partial b_{n-1}}{\partial v} \end{bmatrix}$$

follows the one of section **14.4**. It is not difficult to determine $\boldsymbol{g}$ and $D_{\boldsymbol{g}}$. Comparing the coefficients in (1) and (3) yields

(4) $$b_k = b_{k-1}u + b_{k-2}v + a_k\,.$$

After setting $b_{-1} := b_{-2} := 0$ this contains also the equations

$$b_0 = a_0 \quad \text{and} \quad b_1 = b_0 u + a_1\,.$$

Differentiating (4) with respect to u and v implies further

$$\frac{\partial b_k}{\partial u} = \frac{\partial b_{k-1}}{\partial u}u + \frac{\partial b_{k-2}}{\partial u}v + b_{k-1}\,,$$
$$\frac{\partial b_k}{\partial v} = \frac{\partial b_{k-1}}{\partial v}u + \frac{\partial b_{k-2}}{\partial v}v + b_{k-2}\,.$$

These recurrence relations are analogous to (4). Since $b_{-1} = b_{-2} = 0$, they can be merged into one relation

(5) $$c_k = c_{k-1}u + c_{k-2}v + b_k$$

where

$$c_k := \frac{\partial b_{k+1}}{\partial u} = \frac{\partial b_{k+2}}{\partial v} \text{ for } k = 0, 1, \ldots, n-2, \quad c_{n-1} := \frac{\partial b_n}{\partial u}$$

and in particular $c_{-2} = c_{-1} = 0$.

With these notations the linear system in **14.4** takes on the form

$$\begin{bmatrix} c_{n-1} & c_{n-2} \\ c_{n-2} & c_{n-3} \end{bmatrix} \begin{bmatrix} \Delta u \\ \Delta v \end{bmatrix} + \begin{bmatrix} b_n \\ b_{n-1} \end{bmatrix} = \begin{bmatrix} 0 \\ 0 \end{bmatrix}.$$

The b_k and c_k are determined by the following Horner scheme for quadratic factors.

15.5 The Extended Horner Scheme for Quadratic Factors

When writing the a_k, b_k, c_k underneath each other as in the Horner-scheme

v_0	u_0	a_0	a_1	$\dots$	a_{n-2}	a_{n-1}	a_n
0	0	b_0	b_1	$\dots$	b_{n-2}	$\underline{b_{n-1}}$	$\underline{b_n}$
0	0	c_0	$\dots$	$\underline{c_{n-3}}$	$\underline{c_{n-2}}$	$\underline{c_{n-1}}$	,

one can present the rules (4) and (5) graphically as

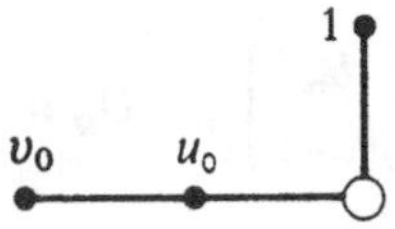

Hence, the algorithm for calculating the b_k and c_k is as follows:

Horner, Quadratic Factors

Input: $a_0, a_1, \dots, a_n;\ u_0, v_0;\ (n \geq 3)$

Output: $b_0, \dots, b_n;\ c_0, \dots, c_{n-1}$

Set $b_{-1} = c_{-2} = c_{-1} := 0;\ b_0 := a_0$.

For $k = 1, 2, \dots, n$

determine $b_k := b_{k-2}v_0 + b_{k-1}u_0 + a_k$

and $c_{k-1} := c_{k-3}v_0 + c_{k-2}u_0 + b_{k-1}$.

Remark 1: The method can be applied to all quadratic factors and is not restricted to those corresponding to pairs of complex conjugate roots.

Example 2: The approximation $\lambda^2 - 0.05\lambda - 3.99$ of the quadratic factor $\lambda^2 - 4$ of

$$\lambda^3 - \lambda^2 - 4\lambda + 4$$

is improved by one Bairstow step. The Horner scheme for the quadratic factor with $u_0 = 0.05$ and $v_0 = 3.99$ is:

3.99	0.05	1	−1	−4	4
0	0	1	−0.95	$\underline{-0.06}$	$\underline{0.21}$
0	0	$\underline{1}$	$\underline{-0.90}$	$\underline{3.88}$	,

and hence the linear system

$$\begin{bmatrix} 3.88 & -0.9 \\ -0.9 & 1 \end{bmatrix} \begin{bmatrix} \Delta u \\ \Delta v \end{bmatrix} + \begin{bmatrix} 0.21 \\ -0.06 \end{bmatrix} = \begin{bmatrix} 0 \\ 0 \end{bmatrix}.$$

The solution $\Delta u = -0.04$, $\Delta v = 0.02$ leads to the improved quadratic factor

$$\lambda^2 - (0.05 - 0.04)\lambda - (3.99 + 0.02) = \lambda^2 - 0.01\lambda - 4.01.$$

It has the approximate zeros $\lambda_1 = 2.01$ and $\lambda_2 = -2.00$.

15.6 Notes and Exercises

1. Changing the coefficient a_k of the polynomial (1) relatively by ε results in the polynomial $\bar{p}(\lambda) = \text{pol}(\lambda) + \varepsilon a_k \lambda^{n-k}$. The simple roots $\bar{\lambda}_i$ of $\bar{p}$ can be determined approximately from the simple roots λ_i of pol(λ). One has

$$\lambda_i - \bar{\lambda}_i \approx \varepsilon \frac{a_k \lambda_i^{n-k}}{\text{pol}'(\lambda_i)}.$$

2. Changing the coefficient $a_1 = 210$ of the polynomial $\text{pol}(\lambda) := \prod_{i=1}^{20} (\lambda - i)$ relatively by ε changes the root $\lambda_{20} = 20$ by $20 - \bar{\lambda}_{20} \approx \varepsilon \cdot 10^{10}$ according to 1 (Wilkinson). Therefore, the roots of this polynomial are ill-conditioned with respect to changes of its coefficients.
3. Splitting of quadratic factors repeatedly by Bairstow's methods distorts particularly the roots computed last. It is advisable to improve these roots finally with the original polynomial.

16 Bernoulli's Method

Newton's method for finding a simple root of a polynomial requires a sufficiently accurate initial approximation. Bernoulli has given a method which is related to the power method to determine such an initial approximation for the root of maximum modulus.

16.1 Linear Difference Equations

The equation

(1) $$a_0 y_{k+n} + a_1 y_{k+n-1} + \cdots + a_{n-1} y_{k+1} + a_n y_k = 0$$

with the constants a_i and $a_0 \neq 0$, $a_n \neq 0$ is called a **homogeneous linear difference equation** of order n with constant coefficients. It defines a sequence $y_n, y_{n+1}, \ldots$ for any choice of the initial values $y_0, \ldots, y_{n-1}$. These sequences are called **solutions** of the linear difference equation.

The polynomial

$$\text{(2)} \qquad \operatorname{pol}(\lambda) := a_0\lambda^n + a_1\lambda^{n-1} + \cdots + a_{n-1}\lambda + a_n$$

with the same coefficients a_i as in (1) is referred to as the **characteristic polynomial** of the linear difference equation (1). One verifies instantly: The sequence $y_k := \lambda_j^k$ of the powers of an arbitrary root λ_j of (2) is a solution for (1) and satisfies $y_{k+1}/y_k = \lambda_j$. Moreover, one gets for an arbitrary solution the following.

Theorem 1: *If (2) has exactly one root λ_1 of maximum modulus, then the quotient y_{k+1}/y_k converges to λ_1 for almost all solutions.*

The proof is given in the next section.

16.2 Matrix Notation

The n consecutive elements $y_{k+1}, \ldots, y_{k+n}$ of a solution of (1) can be obtained from the n consecutive numbers $y_k, \ldots, y_{k+n-1}$. It is expedient here to describe this process with the aid of matrices. Without loss of generality, one may assume $a_0 = 1$; then one gets:

$$\text{(3)} \qquad \begin{bmatrix} 0 & 1 & & \\ & \ddots & \ddots & \\ & & 0 & 1 \\ -a_n & \cdots & -a_2 & -a_1 \end{bmatrix} \begin{bmatrix} y_k \\ \vdots \\ \vdots \\ y_{k+n-1} \end{bmatrix} = \begin{bmatrix} y_{k+1} \\ \vdots \\ \vdots \\ y_{k+n} \end{bmatrix}$$

or for short

$$A\boldsymbol{y}_k = \boldsymbol{y}_{k+1} \text{ with } \boldsymbol{y}_k := [y_k, \ldots, y_{k+n-1}]^t .$$

The following theorem is used to prove Theorem 1.

Theorem 2: *A has the characteristic polynomial*

$$\text{(4)} \qquad \det[A - \lambda I] = (-1)^n(\lambda^n + a_1\lambda^{n-1} + \cdots + a_n) .$$

The proof is by induction on the degree n:

1. For $n = 1$, $[A - \lambda I] = -(\lambda + a_1)$.

2. Assume validity of (4) for degree $n = m - 1$.

3. Then (4) also holds for $n = m$. Namely, if $\det[A - \lambda I]$ for degree m is expanded along the first column, one gets

$$\det \begin{bmatrix} -\lambda & 1 & & & \\ & \ddots & \ddots & & \\ & & -\lambda & 1 & \\ -a_m & \cdots & -a_2 & -a_1 & -\lambda \end{bmatrix} = -\lambda \det \begin{bmatrix} -\lambda & 1 & & \\ & \ddots & \ddots & \\ & & & 1 \\ -a_{m-1} & \cdots & & -a_1 - \lambda \end{bmatrix} - (-1)^{m-1} a_m$$

and because of the induction hypothesis 2,

$$\det[A - \lambda I] = -\lambda(-1)^{m-1}(\lambda^{m-1} + \cdots + a_{m-1}) - (-1)^{m-1} a_m .$$

Hence, A has the characteristic polynomial (4) and its zero of maximum modulus can be determined by the power method provided there is only one such zero.

According to **11.3**, the quotient

$$\frac{y_{k+n}}{y_{k+n-1}} = \frac{e_n^t y_{k+1}}{e_n^t y_k}$$

converges to λ_1 as k increases for all suitable initial vectors y_0. This concludes the proof of Theorem 1.

Remark 1: One can show that, for instance, the initial vector $y_0 := e_n := [0, \ldots, 0, 1]^t$ has a non-vanishing component in the direction of the eigenvector of A associated with the eigenvalue λ_1.

16.3 Bernoulli's Method

The explicit form of the iteration using rule (3) goes back to Bernoulli. The corresponding algorithm is given below:

Bernoulli

Input:	$a_0 \neq 0, a_1, \ldots, a_n; N$
Output:	λ_1 with $\text{pol}(\lambda_1) \approx 0$; $\lvert\lambda_1\rvert > \lvert\lambda_{i \neq 1}\rvert$

Set $y_1 := \cdots := y_{n-1} := 0$; $y_n := 1$.

For $k = 1, 2, \ldots, N$

  determine $y_{k+n} := -\dfrac{1}{a_0}(a_1 y_{k+n-1} + \cdots + a_n y_k)$.

Set $\lambda_1 := \dfrac{y_{N+n}}{y_{N+n-1}}$.

The procedure converges only linearly and with the convergence factor λ_2/λ_1 where λ_1 and λ_2 are the two zeros of (2) of maximum modulus. For this reason, one performs only a few iterations and uses the resulting approximation as the initial value for a faster converging method, e.g., Newton's method.

Example 1: The linear difference equation

$$y_{k+3} - y_{k+2} - 4y_{k+1} + 4y_k = 0$$

has the characteristic polynomial

$$\lambda^3 - \lambda^2 - 4\lambda + 4 .$$

The sequence

$$0,\ 0,\ 1,\ 1,\ 5,\ 5,\ 21,\ 21,\ 85,\ 85,\ \ldots$$

is a solution for the difference equation. The quotients of two consecutive y_k form the sequence

$$\infty,\ 1,\ 5,\ 1,\ 4.2,\ 1,\ 4.05,\ 1,\ \ldots .$$

This sequence does not converge. This example is picked up again in **17.3**.

16.4 Inverse Iteration

On substituting $\lambda = 1/\mu$ in (2) one obtains for $\mu \neq 0$

$$\mathrm{rec}(\mu) := \mu^n \mathrm{pol}\left(\frac{1}{\mu}\right) = a_0 + a_1\mu + \cdots + a_{n-1}\mu^{n-1} + a_n\mu^n$$

which is a polynomial in μ. If $\lambda_n \neq 0$ is the smallest zero of $\mathrm{pol}(\lambda)$ moduluswise, then, $\mu_n := 1/\lambda_n$ is the absolutely largest zero of $\mathrm{rec}(\mu)$ and vice versa. Thus, Bernoulli's method for $\mathrm{rec}(\mu)$ yields the reciprocal value of the absolutely smallest zero of $\mathrm{pol}(\lambda)$.

The respective algorithm is given next.

Bernoulli Inverse

Input: $a_0, a_1, \ldots, a_n \neq 0;\ N$

Output: λ_n with $\mathrm{pol}(\lambda_n) \approx 0;\ |\lambda_n| < |\lambda_{i\neq n}|$

Set $y_1 := \cdots := y_{n-1} := 0;\ y_n := 1$.

For $k = 1, 2, \ldots, N$

determine $y_{k+n} := -\dfrac{1}{a_n}(a_{n-1}y_{k+n-1} + \cdots + a_0 y_k)$.

Set $\lambda_n := \dfrac{y_{N+n-1}}{y_{N+n}}$.

Here too, y_k/y_{k+1} converges only linearly.

Example 2: An approximation for λ_n can be procured also in the following way: On extending the sequence of Example 1 to the left side one gets

$$\ldots, -\frac{21}{64}, -\frac{5}{16}, -\frac{5}{16}, -\frac{1}{4}, -\frac{1}{4}, 0, 0, 1.$$

Two consecutive y_k have the quotients

$$\ldots, \frac{20}{21}, \frac{5}{5}, \frac{4}{5}, \frac{1}{1}, 0.$$

The sequence of these quotients converges to the left towards 1. $\lambda = \frac{20}{21} \approx 0.95$ is a good approximation for the absolutely smallest root $\lambda_3 = 1$. It has been improved by Newton's method in Example 1 of **15.3**.

16.5 Notes and Exercises

1. If the characteristic polynomial (2) of a linear difference equation (1) has a double root $\lambda_i = \lambda_{i+1}$, then the sequence of the $y_k := k\lambda_i^{k-1}$ is also a solution of (1).
2. The linear convergence of the quotients generated by Bernoulli's method can be accelerated by use of Aitken's Δ^2-process **13.4**.
3. If the characteristic polynomial (2) has a pair of complex conjugate numbers $re^{\pm i\varphi}$, $0 < \varphi < \pi$, as its absolutely largest zeros, then the sequence of the solutions $[p_k, q_k]^t$ for the linear systems

$$\begin{bmatrix} y_{k-3} & y_{k-2} \\ y_{k-2} & y_{k-1} \end{bmatrix} \begin{bmatrix} p_k \\ q_k \end{bmatrix} + \begin{bmatrix} y_{k-1} \\ y_k \end{bmatrix} = \begin{bmatrix} 0 \\ 0 \end{bmatrix}$$

converges to $[r^2, -2r\cos\varphi]^t$. The y_k are the elements of a solution of (1).

17 The QD Algorithm

Bernoulli's method yields approximations solely for the absolutely smallest and largest zero of a polynomial. The QD algorithm by Rutishauser (1954) is a generalization of it and determines in certain cases approximations for all roots.

17.1 The LR Algorithm for Tridiagonal Matrices

The LU decomposition of a tridiagonal matrix A whose superdiagonal consists of ones can be written – if its exists – as follows:

$$A = \begin{bmatrix} q_1 & 1 & & \\ e_1 q_1 & e_1{+}q_2 & \ddots & \\ & \ddots & \ddots & 1 \\ & & e_{n-1}q_{n-1} & e_{n-1}{+}q_n \end{bmatrix} = \begin{bmatrix} 1 & & & \\ e_1 & 1 & & \\ & \ddots & \ddots & \\ & & e_{n-1} & 1 \end{bmatrix} \cdot \begin{bmatrix} q_1 & 1 & & \\ & q_2 & \ddots & \\ & & \ddots & 1 \\ & & & q_n \end{bmatrix} = LU\,.$$

The LU algorithm **12.1** which determines the eigenvalues of A multiplies L and U in reverse order,

$$A' = UL = \begin{bmatrix} e_1{+}q_1 & 1 & & \\ e_1 q_2 & e_2{+}q_2 & \ddots & \\ & \ddots & \ddots & 1 \\ & & e_{n-1}q_n & q_n \end{bmatrix},$$

and finds the LU factorization $L'U'$ of the product if it exists. On comparing the product UL with its factorization $L'U'$ one gets the formulas

$$e'_{k-1} + q'_k = e_k + q_k\,, \quad e'_k \cdot q'_k = e_k \cdot q_{k+1}$$

where $e'_0 = e'_n = 0$ for all $k = 1, 2, \ldots, n$.

Often, these rules are graphically depicted as **rhombus rules**:

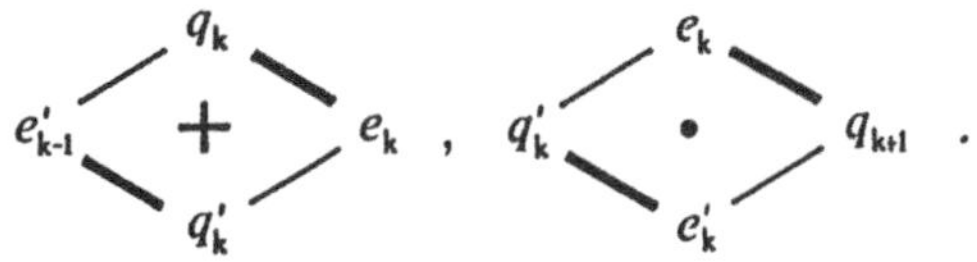

The sums (products) of the elements connected by a fat line are equal.

These rhombus rules allow for an organized computation of the LU decomposition $L'U'$ if the elements e_k, q_k of L and U are arrayed along an oblique line.

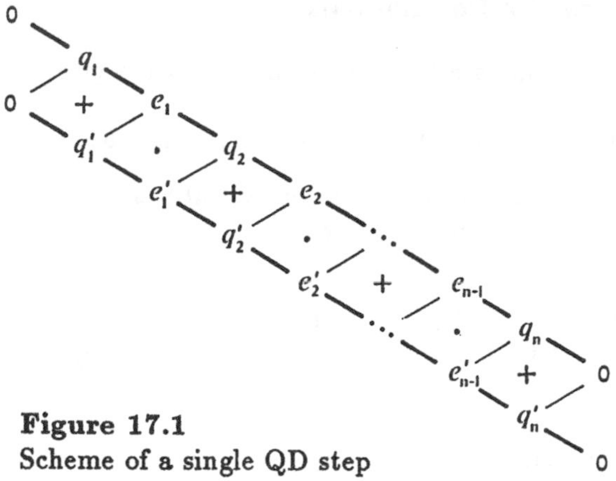

Figure 17.1
Scheme of a single QD step

The corresponding algorithm is as follows:

QD Step

Input:	$e_1, \ldots, e_{n-1};\ q_1, \ldots, q_n$
Output:	$e'_1, \ldots, e'_{n-1};\ q'_1, \ldots, q'_n$

Set $e_0 := 0$.
For $k = 1, 2, \ldots, n-1$
  $q_k := q_k + e_k - e_{k-1}$.
  If $q_k = 0$, then **stop**,
  else $e_k := e_k \dfrac{q_{k+1}}{q_k}$.
$q_n := q_n - e_{n-1}$.

Note: The algorithm fails if some q_k, $k = 1, 2, \ldots, n-1$ is zero. The algorithm is also called **quotient difference algorithm** because of the respective formulas for e_k and q_k. One can infer directly from **12.2** the following

Theorem 1: *Let the moduli of the eigenvalues of A be pairwise distinct, e.g.,*

$$|\lambda_1| > |\lambda_2| > \cdots > |\lambda_n| .$$

Then, on repeating the QD step, the e_k converge to zero while the q_k converge to the (in general ordered) eigenvalues of A.

The corresponding scheme comprising all QD steps is termed **QD scheme**.

17.2 The QD Scheme for Polynomials

There are several ways of building a tridiagonal matrix with a given characteristic polynomial

$$\mathrm{pol}(\lambda) = a_0\lambda^n + a_1\lambda^{n-1} + \cdots + a_n\,, \quad a_0 \neq 0\,.$$

An especially simple construction has been given by Rutishauser for the case that all $a_i \neq 0$. Beginning with a horizontal zigzag row

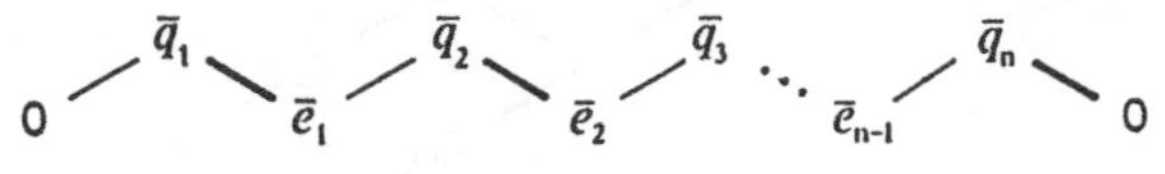

containing the initial numbers

$$\bar{q}_1 := -\frac{a_1}{a_0}\,, \text{ and } \bar{q}_2 = \cdots = \bar{q}_n := 0\,, \text{ and}$$
$$\bar{e}_1 := \frac{a_2}{a_1}\,, \quad \ldots\,, \quad \bar{e}_{n-1} := \frac{a_n}{a_{n-1}}$$

oblique rows are produced from the right to the left using the rhombus rules:

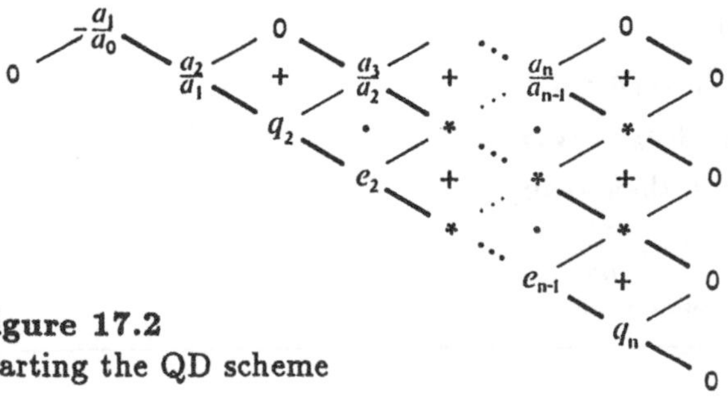

Figure 17.2
Starting the QD scheme

Then, the following theorem applies whose lengthy proof is omitted here.

Theorem 2: *The tridiagonal matrix LU which belongs to the leftmost oblique row has the given polynomial as its characteristic polynomial.*

The corresponding algorithm is formulated below:

Start QD Scheme

Input: $a_0 \neq 0, \ldots, a_n \neq 0$

Output: $e_1, \ldots, e_{n-1};\ q_1, \ldots, q_n$

```
Set e_0 := e_n := q_2 := q_3 := ... := q_n := 0; q_1 := -a_1/a_0 .
For i = n, n-1, ..., 2
    set e_{i-1} := a_i/a_{i-1} and
    for k = i, i+1, ..., n-1
        q_k := q_k + e_k - e_{k-1} .
        If q_k = 0, then failure,
        else e_k := e_k q_{k+1}/q_k .
q_n := q_n - e_{n-1} .
```

Remark 1: One can show that with this start the first column of the QD scheme coincides with the sequence of the quotients formed by Bernoulli's method when using the initial vector $[0, \ldots, 0, 1]^t$.

17.3 Pairs of Zeros with Equal Modulus

Consider the case that A has absolutely equal eigenvalues, for instance $|\lambda_{j+1}| = |\lambda_j|$. Then, according to **12.3**, A' does not converge to an upper triangular matrix and the e_j of the jth column of the QD scheme do not converge to zero. However, if λ_j and λ_{j+1} are isolated, i.e.,

$$\cdots \geq |\lambda_{j-1}| > |\lambda_j| = |\lambda_{j+1}| > |\lambda_{j+2}| \geq \cdots ,$$

then the e_j, $i \neq j$, go to zero, cf. **12.3**, while the submatrices

$$Q := \begin{bmatrix} q_j & 1 \\ e_j q_j & e_j + q_{j+1} \end{bmatrix}$$

diverge, in general. The eigenvalues of the matrices Q, however, converge to λ_j and λ_{j+1}. Therefore

$$\det[Q - \lambda I] = \lambda^2 - (q_j + e_j + q_{j+1})\lambda + q_j q_{j+1}$$

is an approximation for the quadratic factor $(\lambda - \lambda_j)(\lambda - \lambda_{j+1})$ of the characteristic polynomial of A. The coefficients can be easily read off the last oblique row of the QD scheme.

Example 1: The QD scheme for the polynomial

$$\lambda^3 - \lambda^2 - 4\lambda + 4$$

with Rutishauser's start has the form

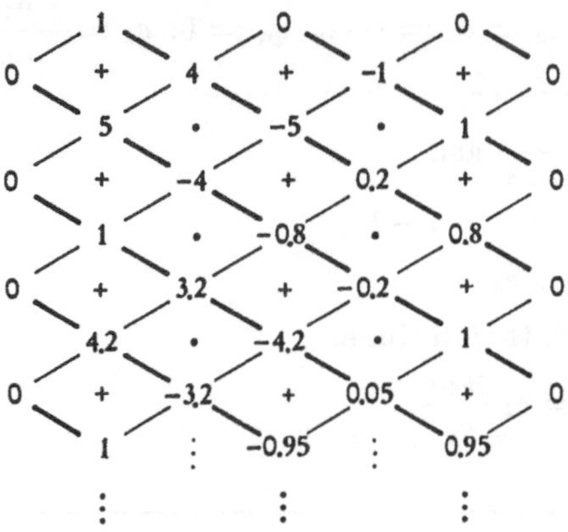

It is more convenient to compute vertical than oblique rows. Obviously, e_2 converges to zero, but not so e_1. Hence, $\lambda - 0.95$ is an approximation for a linear factor of the polynomial and $\lambda^2 - (4.2 - 3.2 - 0.95)\lambda - 0.95 \cdot 4.2$ for a quadratic factor. They have been improved already in **15.3** and **15.5** by the methods of Newton and Bairstow.

The quotients in the first column of the QD scheme equal the quotients computed in Example 1 of **16.3** using Bernoulli's method.

17.4 Notes and Exercises

1. It is possible to build the QD scheme from its first column which can be computed by Bernoulli's method. However, this is a highly unstable numerical procedure where the e_n fail, in general, to vanish.
2. The sums of all q_i in each row equal $-a_1/a_0$. One can use this fact to check the results.
3. The products of all q_i in each oblique row equal $(-1)^n a_n/a_0$.
4. The QD scheme can be used analogously to determine the approximations for the zeros of power series. To this end, one has to compute successively the oblique rows rising from the left to the right.

IV Interpolation and Discrete Approximation

An important task is to approximate a complicated function $f(x)$ by another function $a(x)$ which is simpler. One way of solving this problem is by interpolating f at discrete values.

18 Interpolation

A function f in one or more variables is often given by a table:

x_0	x_1	$\cdots$	x_n
f_0	f_1	$\cdots$	f_n ,

	y_0	$\cdots$	y_m
x_0	$f_{0,0}$	$\cdots$	$f_{0,m}$
$\vdots$	$\vdots$		$\vdots$
x_n	$f_{n,0}$	$\cdots$	$f_{n,m}$.

In order to obtain the intermediate values of f which are not tabulated one has to **interpolate** between neighboring values of the table.

18.1 Interpolation Polynomials

The interpolation problem in one variable can be geometrically stated as follows:

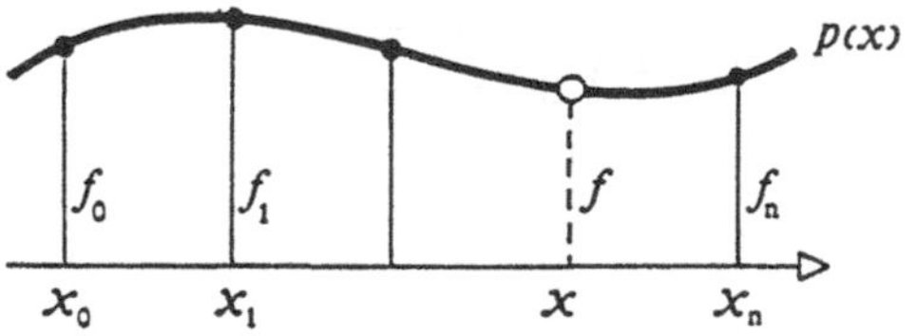

Figure 18.1
Interpolation polynomial

Given $n+1$ distinct points with the coordinates x_i, f_i, one is to find a curve $p(x)$ passing through these points and evaluate the curve for some argument x. The x_i are supposed to be distinct and are called **interpolation abscissae.** The corresponding f_i are called **interpolation ordinates** and the corresponding points (x_i, f_i) **interpolation points.**

When interpolating with polynomials, one has

Theorem 1: *If $x_0, \ldots, x_n$ are $n+1$ distinct arguments with corresponding ordinates $f_0, \ldots, f_n$, then there exists a unique polynomial* pol(x) *of degree at most n with the property that*

$$\mathrm{pol}(x_i) = f_i\,, \qquad i = 0, \ldots, n\,.$$

For the proof, one writes a polynomial of degree n in the form[1)]

$$\text{(1)} \qquad \mathrm{pol}(x) = c_0 + c_1 x + \cdots + c_{n-1} x^{n-1} + c_n x^n .$$

Successively inserting all $n+1$ interpolation points produces the linear system

$$\begin{bmatrix} 1 & x_0 & \cdots & x_0^n \\ \vdots & \vdots & & \vdots \\ 1 & x_n & \cdots & x_n^n \end{bmatrix} \begin{bmatrix} c_0 \\ \vdots \\ c_n \end{bmatrix} = \begin{bmatrix} f_0 \\ \vdots \\ f_n \end{bmatrix}$$

for the c_k. Its matrix is the so-called **Vandermonde matrix** which is non-singular for pairwise distinct x_i. The reason for this is that no polynomial of degree $\le n$ can have more than n zeros unless it is identical zero, i.e., the homogeneous system obtained for $f_0 = \cdots = f_n = 0$ has only the trivial solution. This entails that the c_k are uniquely determined for any choice of the f_i. Some c_k, in particular c_n, may be zero, though. E.g., if $f_0 = f_1 = \cdots f_n = 1$, then $c_0 = 1$, $c_1 = \cdots = c_n = 0$.

18.2 Lagrange Polynomials

It is a well-known fact that all polynomials of degree n or less form a vector space of dimension $n+1$. The polynomials $1, x, x^2, \ldots, x^n$ used in expression (1) form a basis for this space. Lagrange introduced a different basis which depends on the interpolation abscissae x_i, namely the $n+1$ polynomials $l_i(x)$ which assume the value 1 at x_i and vanish at all the other n support abscissae. Hence, they satisfy

$$l_i(x_k) = \delta_{i,k} := \begin{cases} 1 & \text{if } i = k \\ 0 & \text{if } i \ne k \,. \end{cases}$$

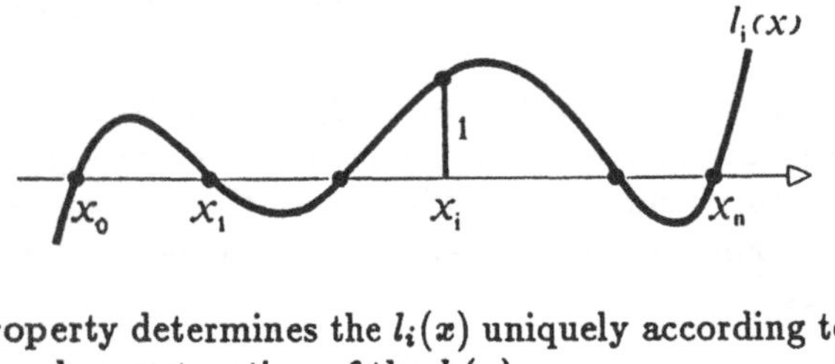

Figure 18.2
Lagrange polynomial

This property determines the $l_i(x)$ uniquely according to Theorem 1. It also allows for a simple construction of the $l_i(x)$:

Because all its zeros are given, $l_i(x)$ has the factorization

$$l_i(x) = a_i(x - x_0) \cdots (x - x_{i-1})(x - x_{i+1}) \cdots (x - x_n)$$

[1)] The indices differ from those in **15**.

with an appropriate constant a_i. The stipulation $l_i(x_i) = 1$, then, determines a_i, i.e.,

$$a_i = \frac{1}{(x_i - x_0)\cdots(x_i - x_{i-1})(x_i - x_{i+1})\cdots(x_i - x_n)}.$$

Remark 1: These expressions become easier in case the abscissae x_i are equidistant, i.e.,

$$x_i := x_0 + ih.$$

Then, introducing the new variable s with

$$x = x_0 + sh$$

yields

$$\frac{x - x_k}{x_i - x_k} = \frac{(x_0 + sh) - (x_0 + kh)}{(x_0 + ih) - (x_0 + kh)} = \frac{s - k}{i - k}$$

and simplifies, thus, the **Lagrange polynomial** $l_i(x)$ to

$$l_i(x(s)) = \frac{(s-0)\cdots(s-i+1)(s-i-1)\cdots(s-n)}{(i-0)\cdots(1)(-1)\cdots(i-n)}.$$

18.3 Lagrange Form

The interpolation polynomial is now readily obtained as

(2) $$f(x) := \mathrm{pol}(x) = f_0 l_0(x) + \cdots + f_n l_n(x).$$

This is verified by inserting the interpolation points, namely

$$\mathrm{pol}(x_k) = \sum_{i=0}^{n} f_i\, l_i(x_k) = \sum_{i=0}^{n} f_i\, \delta_{i,k} = f_k.$$

In order to evaluate (2) one extracts from all $l_i(x)$ the product

$$a := (x - x_0)\cdots(x - x_n),$$

namely

$$l_i(x) = a\frac{a_i}{x - x_i}, \qquad x \neq x_i.$$

Using the abbreviations $q_i := \dfrac{a_i}{x - x_i}$ one can write (2) then as

$$f = \mathrm{pol}(x) = a(f_0 q_0 + \cdots + f_n q_n).$$

Because a and the q_i are the same for all values of the f_i, one can assume $f_0 = \cdots = f_n = 1$ to determine a. Then $f = 1$ and therefore

$$a = \frac{1}{q_0 + \cdots + q_n} =: \frac{1}{q}.$$

Thus $l_i(x) = q_i/q$ and

$$f = \frac{f_0 q_0 + \cdots + f_n q_n}{q}.$$

This means, f is the **weighted average** of the f_i. The algorithm is as follows:

Lagrange

Input: $x_0, \ldots, x_n, x$ distinct; $f_0, \ldots, f_n$

Output: $f = \mathrm{pol}(x)$

1 For $i = 0, 1, \ldots, n$

2 set $q_i := 1$,

3 for $k = 0, 1, \ldots, n$; $k \neq i$

4 set $q_i := q_i(x_i - x_k)$.

5 Set $q_i := \dfrac{1}{q_i(x - x_i)}$.

6 Determine $f := \dfrac{f_0 q_0 + \cdots + f_n q_n}{q_0 + \cdots + q_n}$.

The instructions **2**, **3** and **4** determine $1/a_i$ which, for the sake of simplicity, is called q_i.

Example 1: The function f given by the table

$x_i =$	0	1	2
$f_i =$	8	5	4

is extrapolated at the argument $x = 3$: Expediently, $x - x_i$ is written $-(x_i - x)$. Then one gets

$$q_0 = \frac{-1}{(0-1)(0-2)(0-3)} = +\frac{1}{6},$$
$$q_1 = \frac{-1}{(1-0)(1-2)(1-3)} = -\frac{3}{6},$$
$$q_2 = \frac{-1}{(2-0)(2-1)(2-3)} = +\frac{3}{6},$$

and $q = \frac{1}{6} - \frac{3}{6} + \frac{3}{6} = \frac{1}{6}$ and finally $f = 8 \cdot 1 - 5 \cdot 3 + 4 \cdot 3 = 5$.

18.4 Newton Form

Another basis, also depending on the interpolation abscissae, has been introduced by Newton, namely the $n+1$ monic polynomials $n_i(x)$ of degree i which vanish at the first i interpolation abscissae:

$$n_i(x) := (x - x_0) \cdots (x - x_{i-1}) .$$

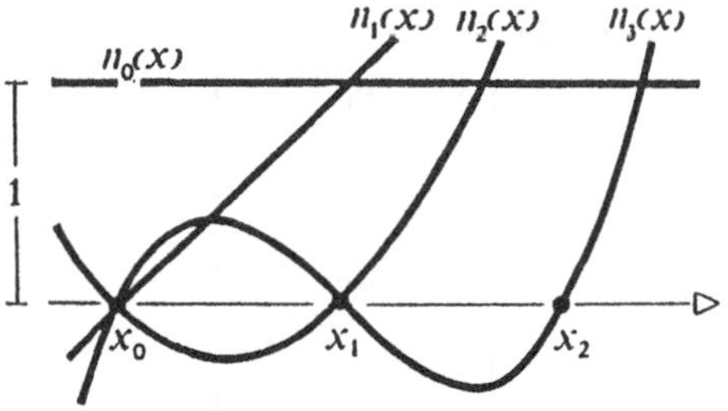

Figure 18.3
Newton polynomials

The representation of the interpolation polynomial (1) in terms of these polynomials is

$$\text{(3)} \qquad \text{pol}(x) = f_0 n_0(x) + f_{0,1} n_1(x) + \cdots + f_{0,\ldots,n} n_n(x) .$$

The $f_{0,\ldots,i}$ are called **divided differences.** They satisfy the recurrence relation

$$f_{i,\ldots,k} = \frac{f_{i,\ldots,k-1} - f_{i+1,\ldots,k}}{x_i - x_k} ;$$

cf. **18.12** for a proof of it. The calculation of the $f_{0,\ldots,i}$ using this formula is effected best in **Newton's scheme:**

	x_0	f_0					
i	$\underline{x_1}$	f_1	$f_{0,1}$				
	x_2	f_2	$\underline{f_{1,2}}$	$f_{0,1,2}$			
k	$\underline{x_3}$	f_3	$\underline{f_{2,3}}$	$\underline{\underline{f_{1,2,3}}}$	$f_{0,1,2,3}$		
	$\vdots$	$\vdots$	$\vdots$			$\ddots$	
	x_n	f_n	$f_{n-1,n}$	$\cdots$		$\cdots$	$f_{0,1,\ldots,n}$.

The values involved in the computation of $f_{1,2,3}$ are underlined as an example.

The advantage of the Newton representation is that the first $i+1$ terms of (3) interpolate the first $i+1$ support points. The degree of the interpolation polynomial can, therefore, easily be increased or decreased.

Example 2: For the problem of Example 1 one obtains

$$n_0 = 1 , \quad n_1(3) = (3 - 0) = 3 , \quad n_2(3) = (3 - 0)(3 - 1) = 6 .$$

The divided difference table is

$$\begin{array}{c|ccc} 0 & 8 & & \\ 1 & 5 & -3 & \\ 2 & 4 & -1 & 1. \end{array}$$

Thus, $f = 8 \cdot 1 - 3 \cdot 3 + 1 \cdot 6 = 5$ is the result.

Remark 2: A comparison of the equations (1) and (3) shows that $f_{0,\dots,n} = c_n$. By means of Cramer's rule one can conclude

$$f_{0,\dots,n} = c_n = \frac{\det \begin{bmatrix} 1 & \cdots x_0^{n-1} & f_1 \\ \vdots & \vdots & \vdots \\ 1 & \cdots x_n^{n-1} & f_n \end{bmatrix}}{\det \begin{bmatrix} 1 & \cdots & x_0^n \\ \vdots & & \vdots \\ 1 & \cdots & x_n^n \end{bmatrix}}.$$

Remark 3: Moreover, $f_{i,\dots,k}$ is seen to be the leading coefficient of the partial interpolation polynomial $p_{i,\dots,k}$ (it may vanish). The notation $p_{i,\dots,k}$ is defined below in the next section.

18.5 Aitken's Lemma

The unique polynomial interpolating just a selection $i, \dots, k$ of the interpolation points shall be denoted by $p_{i,\dots,k}(x)$, for the sake of simplicity.

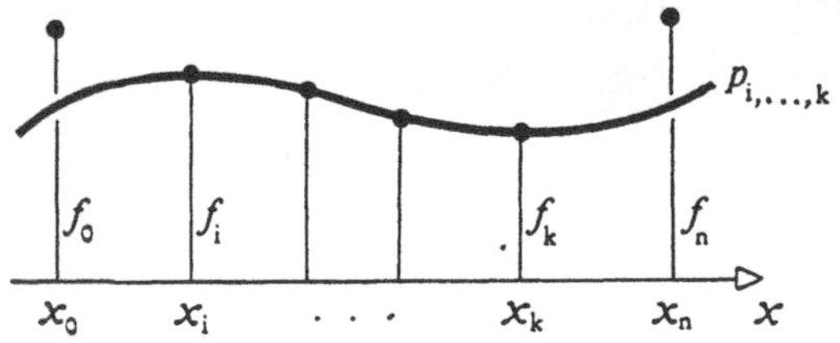

Figure 18.4
Partial interpolation

The notation $p_{i,\dots,k}$ does not depend on the order of the interpolation points and their indices. Special cases are

$$f_i = p_i\,, \quad \text{and} \quad \mathrm{pol}(x) = p_{0,\dots,n}(x)\,.$$

For $n = 1$ one obtains from (2) the well-known formula

$$f = f_0 \frac{x - x_1}{x_0 - x_1} + f_1 \frac{x - x_0}{x_1 - x_0}$$

for the **linear interpolation** by the polynomial $p_{0,1}$.

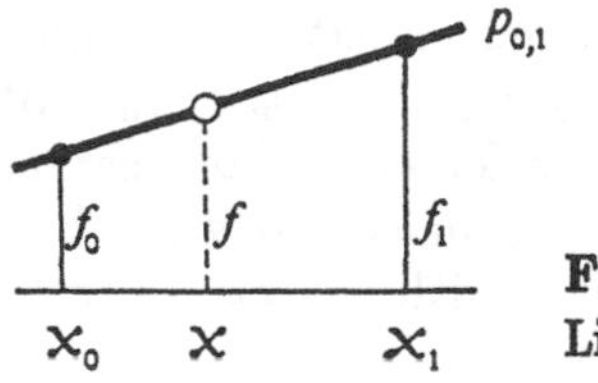

Figure 18.5
Linear interpolation

This formula can be used to construct $p_{i,\dots,k}$. One can assume without loss of generality that $x_i, x_{i+1}, x_{i+2}, \dots, x_k$ are corresponding interpolation abscissae. Then $p_{i,\dots,k-1}$ and $p_{i+1,\dots,k}$ are two different polynomials and each has $k-i$ interpolation points in common with $p_{i,\dots,k}$. The polynomials satisfy the following relation.

Aitken's Lemma: *$p_{i,\dots,k}(x)$ is obtained from $p_{i,\dots,k-1}(x)$ and $p_{i+1,\dots,k}(x)$ by linear interpolation. Namely $p_{i,\dots,k}(x) = p(x)$ with*

$$p(x) = \frac{(x_k - x)p_{i,\dots,k-1}(x) - (x_i - x)p_{i+1,\dots,k}(x)}{x_k - x_i}.$$

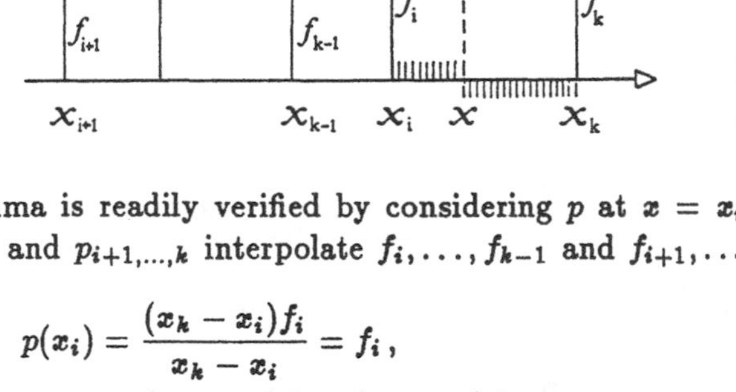

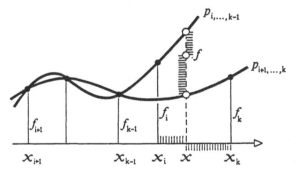

Figure 18.6
Aitken's Lemma

The lemma is readily verified by considering p at $x = x_i, \dots, x = x_k$. Because $p_{i,\dots,k-1}$ and $p_{i+1,\dots,k}$ interpolate $f_i, \dots, f_{k-1}$ and $f_{i+1}, \dots, f_k$ respectively, it follows

$$p(x_i) = \frac{(x_k - x_i)f_i}{x_k - x_i} = f_i\,,$$

$$p(x_j) = \frac{(x_k - x_j)f_j - (x_i - x_j)f_j}{x_k - x_i} = f_j\,, \quad j = i+1, \dots, k-1\,,$$

$$p(x_k) = \frac{-(x_i - x_k)f_k}{x_k - x_i} = f_k\,.$$

The degrees of $p_{i,\dots,k-1}$ and $p_{i+1,\dots,k}$ are less than $k-i$. Hence p is of degree $k-i$ at most and therefore agrees with the unique interpolation polynomial $p_{i,\dots,k}$ according to Theorem 1.

18.6 Neville's Scheme

An interpolation polynomial $p_{0,\ldots,n}(x)$ and, in particular, its value at some numbers x can be constructed with the aid of Aitken's Lemma by **iterated linear interpolation** from the constants $p_0 = f_0, \ldots, p_n = f_n$. Aitken, Neville and others have introduced different schemes for this. For example, Neville determines all interpolation polynomials $p_{i,i+1,\ldots,k}$ successively for $k - i = 1, 2, \ldots, n$. It is convenient to array the $p_{i,\ldots,k}$ together with the $x_k - x$ in **Neville's scheme**:

$$
\begin{array}{cc|ccccc}
 & x_0 - x & p_0 & & & & \\
i & \underline{x_1 - x} & p_1 & p_{0,1} & & & \\
 & x_2 - x & p_2 & \underline{p_{1,2}} & p_{0,1,2} & & \\
k & \underline{x_3 - x} & p_3 & \underline{p_{2,3}} & \underline{\underline{p_{1,2,3}}} & p_{0,1,2,3} & \\
 & \vdots & \vdots & \vdots & & & \ddots \\
 & x_n - x & p_n & p_{n-1,n} & \cdots & \cdots & p_{0,1,\ldots,n}\,.
\end{array}
$$

The $p_{i,\ldots,k}$ are computed by means of Aitken's Lemma row after row or column after column. The values involved for calculating $p_{1,2,3}$ are underlined as an example. The algorithm below determines the columns from the left to the right and writes them over the first one:

Neville

Input: $x_0, \ldots, x_n$ distinct; $f_0, \ldots, f_n$; x

Output: $p_n := p_{0,\ldots,n}$

```
For k = 0, 1, ..., n
    set p_k := f_k .
For j = 1, 2, ..., n
    and for k = n, n-1, ..., j
        determine p_k := ((x_k - x) p_{k-1} - (x_{k-j} - x) p_k) / (x_k - x_{k-j}) .
```

$$p_k := \frac{(x_k - x)p_{k-1} - (x_{k-j} - x)p_k}{x_k - x_{k-j}}\,.$$

Remark 4: Here, $x \neq x_i$ is not required.

Example 3: The following is Neville's scheme for the problem of Example 1.

$$
\begin{array}{c|ccc}
0-3 & 8 & & \\
1-3 & 5 & -1 & \\
2-3 & 4 & 3 & 5\,.
\end{array}
$$

Hence $f = p_{0,1,2}(3) = 5$.

18.7 Hermite Interpolation

In **18.1** polynomials are considered which interpolate certain values of a function f. More general, one can construct polynomials that interpolate values of f and of its derivatives $f^{(k)}$.

Interpolation of a prescribed number of consecutive derivatives at the interpolation abscissae is called **Hermite interpolation.**

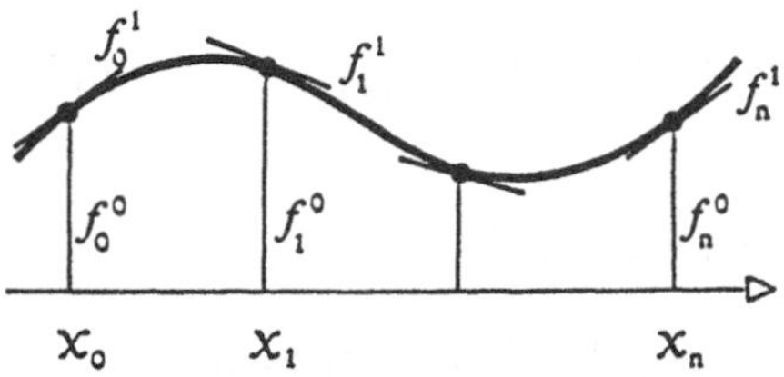

Figure 18.7
Hermite interpolation

The existence of interpolating **Hermite polynomials** is provided by a theorem which generalizes Theorem 1.

Theorem 2: *Let $x_0, \ldots, x_n$ be $n+1$ distinct interpolation abscissae and $m_0, \ldots, m_n$ some associated non-negative integers. Then there exists a unique interpolation polynomial* pol(x) *of degree $m = n + m_0 + \cdots + m_n$ which solves*

$$\mathrm{pol}^{(k)}(x_i) = f_i^k\,,$$

$k = 0, 1, \ldots, m_i$ and $i = 0, 1, \ldots, n$, for any given f_i^k.

The proof is similar to the one of Theorem 1. Namely on writing the interpolation polynomial in the form

$$\mathrm{pol}(x) = c_0 + c_1 x + \cdots + c_m x^m$$

one finds that the interpolation constraints form a linear system with $m+1$ equations for the $m+1$ unknowns $c_0, \ldots, c_m$. For the solution of the associate homogeneous system,

$$\mathrm{pol}^{(k)}(x_i) = 0\,,$$

it is well-known that pol(x) has an m_i-fold root at x_i for all $i = 0, \ldots, n$. This means altogether, that pol(x) has $m+1$ roots but is of degree only m, i.e., the homogeneous system has only the trivial solution which implies that the inhomogeneous system, for all choices of the f_i^k, has always a unique solution.

18.8 Piecewise Hermite Interpolation

The general Hermite polynomials in **18.7** are rather cumbersome for large m. Therefore one resorts to **piecewise Hermite interpolation**, i.e., one solves separate interpolation problems for each interval $[x_i, x_{i+1}]$, $i = 0, \ldots, n-1$.

For piecewise Hermite interpolation it is necessary to set $m_0 = \cdots = m_n = m$. Then one has to find the n unique Hermite polynomials $\mathrm{pol}_i(x)$, $i = 0, \ldots, n-1$, of degree $r = 2m+1$ which solve the interpolation problem

$$(4) \qquad \mathrm{pol}_i^{(k)}(x_i) = f_i^k\,, \qquad \mathrm{pol}_i^{(k)}(x_i+1) = f_{i+1}^k\,, \qquad k = 0, 1, \ldots, m\,.$$

Remark 5: The piecewise polynomial function

$$s(x) = \mathrm{pol}_i(x) \quad \text{if} \quad x \in [x_i, x_{i+1}]$$

is m-times differentiable in $[x_0, x_n]$. Such a function is known as a **spline**, cf. **21**.

18.9 The Cardinal Hermite Basis

For the further discussion of the interpolation problem (4) it suffices to consider the standard interval [0,1] of a **local parameter** λ. Then the interpolation constraints are of the form

$$(5) \qquad \mathrm{pol}^{(k)}(0) = f_0^k\,, \qquad \mathrm{pol}^{(k)}(1) = f_1^k\,, \qquad k = 0, 1, \ldots, m\,.$$

The Hermite polynomials H_i^l, $i = 0, 1$, $l = 0, 1, \ldots, m$ of degree $r = 2m+1$ that solve (5) for

$$f_j^k = \delta_{i,j}\,\delta_{k,l}$$

are the **cardinal Hermite basis functions** of degree r.

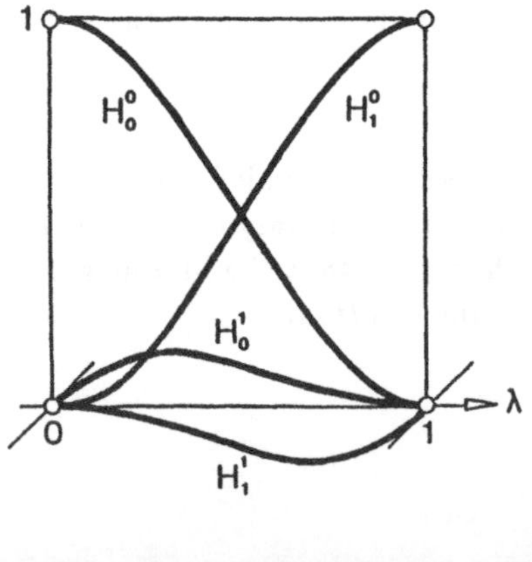

Figure 18.8
Cardinal cubic Hermite functions

One can find the cardinal Hermite basis functions $H_i^l(x) = a_0 + a_1\lambda + \cdots + a_r\lambda^r$ by solving the corresponding interpolation equations for the coefficients $a_0, \ldots, a_r$. A simpler approach is described in **20.5.**

Example 4: The cubic cardinal Hermite functions are

$$H_0^0(\lambda) = 1 - 3\lambda^2 + 2\lambda^3 \qquad H_0^1(\lambda) = \lambda - 2\lambda^2 + \lambda^3$$

and symmetrically

$$H_1^0(\lambda) = H_0^0(1-\lambda) \qquad H_1^1(\lambda) = -H_0^1(1-\lambda)\,.$$

The solution for the general problem (5) is now easily obtained as

$$\mathrm{pol}(\lambda) = \sum_{k=0}^{1} f_0^k\, H_0^k(\lambda) + \sum_{k=0}^{1} f_1^k\, H_1^k(\lambda)\,.$$

Example 5: The more general Hermite interpolation problem

$$\begin{aligned} \mathrm{pol}(x_0) &= 7\,, & \mathrm{pol}(x_1) &= 3\,, \\ \mathrm{pol}'(x_0) &= 2\,, & \mathrm{pol}'(x_1) &= 4 \end{aligned}$$

has the unique cubic solution

$$\mathrm{pol}(\lambda) = 7 \cdot H_0^0(\lambda) + 2 \cdot \Delta x_0 H_0^1(\lambda) + 3 \cdot H_1^0(\lambda) + 4 \cdot \Delta x_0 H_1^1(\lambda)\,,$$

where the local coordinate λ is defined by $x = x_0(1-\lambda) + x_1\lambda$. Note that $\Delta x_0 \cdot \mathrm{pol}'(x) = \frac{d}{d\lambda}\mathrm{pol}(\lambda)$.

18.10 More-Dimensional Interpolation

The interpolation problem in two variables can be formulated analogously. Given $(n+1)(m+1)$ points $(x_i, y_k, f_{i,k})$, $i = 0, \ldots, n$; $k = 0, \ldots, m$; a surface $f(x, y)$ passing through these points is to be found and evaluated at some point (x, y).

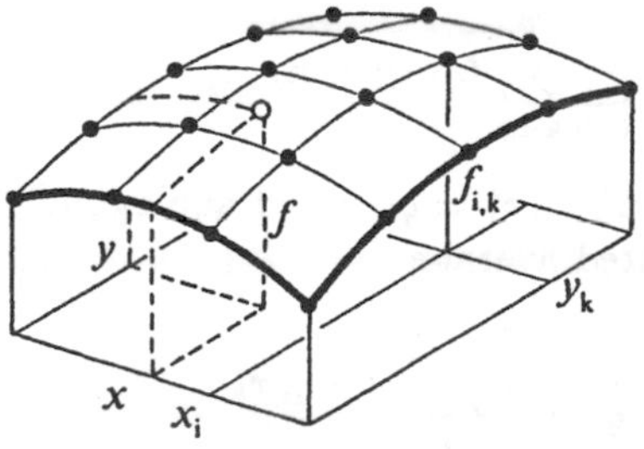

Figure 18.9
Interpolation surface

In analogy to one-dimensional interpolation, bivariate polynomials can be used to fit those points:

$$f(x,y) := \text{pol}(x,y) := \sum_{r=0}^{n}\sum_{s=0}^{m} c_{r,s}\, x^r y^s .$$

Inserting the interpolation points into this equation results in a system of linear equations for the $(n+1)(m+1)$ coefficients $c_{r,s}$, which can be written in terms of matrices:

$$\begin{bmatrix} 1 & x_0 & \cdots & x_0^n \\ \vdots & \vdots & & \vdots \\ 1 & x_n & \cdots & x_n^n \end{bmatrix} \begin{bmatrix} c_{0,0} & \cdots & c_{0,m} \\ \vdots & & \vdots \\ c_{n,0} & \cdots & c_{n,m} \end{bmatrix} \begin{bmatrix} 1 & \cdots & 1 \\ y_0 & \cdots & y_m \\ \vdots & & \vdots \\ y_0^m & \cdots & y_m^m \end{bmatrix} = \begin{bmatrix} f_{0,0} & \cdots & f_{0,m} \\ \vdots & & \vdots \\ f_{n,0} & \cdots & f_{n,m} \end{bmatrix} .$$

It has a unique solution provided the interpolation abscissae x_i and the interpolation abscissae y_k are pairwise distinct.

However, for practical applications, other basis representations of pol(x,y) are preferable, for instance the bivariate polynomials $l_{i,k}(x,y)$ which vanish at all interpolation abscissae but one where they assume the value 1, i.e.,

$$l_{i,k}(x_r, y_s) = \delta_{i,r}\, \delta_{k,s} . \tag{6}$$

They can be obtained easily. Let $l_i(x)$ and $m_k(y)$ be the Lagrange polynomials associated with the interpolation abscissae x_i and y_k respectively. Then,

$$l_{i,k}(x,y) := l_i(x)\, m_k(y)$$

satisfies the condition (6).

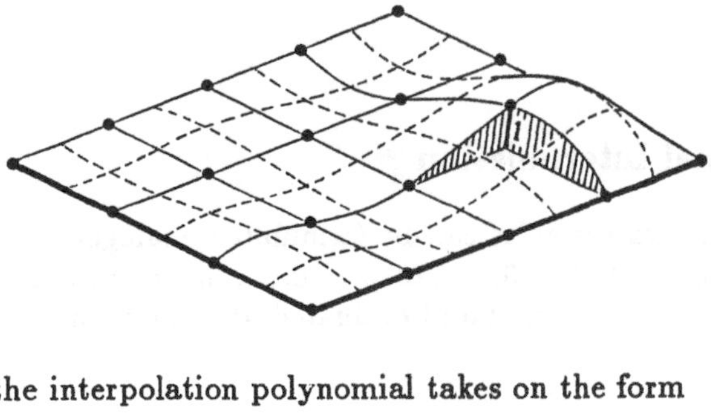

Figure 18.10
Lagrange surface

Now, the interpolation polynomial takes on the form

$$\text{pol}(x,y) = \sum_{i=0}^{n}\sum_{k=0}^{m} f_{i,k}\, l_i(x) m_k(y)$$

which can be evaluated like (2). With $l_i(x) = q_i/q$ and analogously $m_k(y) = r_k/r$ one can write pol(x,y) as the weighted average

$$\text{pol}(x,y) = \frac{1}{q \cdot r} \sum_{i=0}^{n}\sum_{k=0}^{m} f_{i,k}\, q_i\, r_k .$$

18.11 Surface Patches of Coons and Gordon

1967, **Coons** introduced a different method for the two-dimensional interpolation which was generalized by Gordon in 1969. Coons and Gordon interpolate the net of curves $f(x_i, y)$, $i = 0, \ldots, n$, and $f(x, y_j)$, $j = 0, \ldots, m$ of some imaginary surface $f(x, y)$ using a one-dimensional interpolation scheme P to construct the interpolation surfaces

$P_x f$ which interpolates the points $f(x_0, y), \ldots, f(x_n, y)$ for every $y =$ constant,

$P_y f$ which interpolates the points $f(x, y_0), \ldots, f(x, y_m)$ for every $x =$ constant.

In particular, the surfaces $P_x f$ and $P_y f$ interpolate the points $f(x_0, y_0), \ldots, f(x_n, y_m)$. If P is a linear interpolation scheme, one can show that

$$P_x P_y f = P_y P_x f\,.$$

Coons and Gordon blend these three interpolation surfaces together by forming the **Boolean sum**

$$Qf = (P_x \oplus P_y)f = (P_x + P_y - P_x P_y)f\,.$$

It is not difficult to check that indeed Qf interpolates the curves $f(x_i, y)$ and $f(x, y_j)$ of f.

Various P can serve as interpolation schemes: Lagrange or Hermite interpolants or even Taylor polynomials, etc., Figure 18.11 shows some examples.

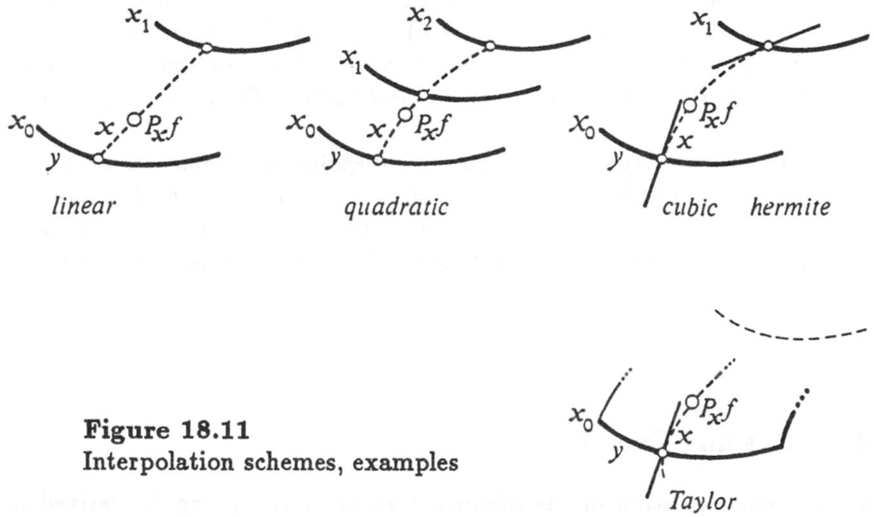

Figure 18.11
Interpolation schemes, examples

Example 6: Figure 18.12 depicts the simple case of constructing a Coons patch where opposite boundary points are interpolated linearly.

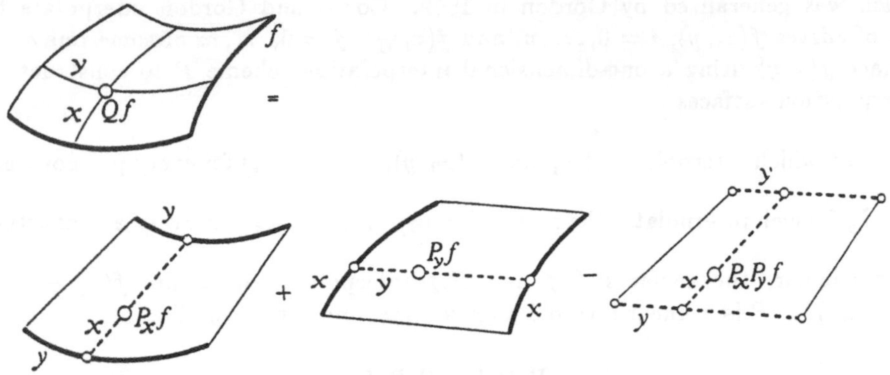

Figure 18.12 Boolean sum of linear interpolants

Using the linear Lagrange polynomials $l_i(x)$ and $m_j(y)$, this Coons patch can be presented by a matrix:

$$Qf = -[-1,\ l_0(x),\ l_1(x)] \left[\begin{array}{c|cc} 0 & f(x,y_0) & f(x,y_1) \\ \hline f(x_0,y) & f(x_0,y_0) & f(x_0,y_1) \\ f(x_1,y) & f(x_1,y_0) & f(x_1,y_1) \end{array}\right] \begin{bmatrix} -1 \\ m_0(y) \\ m_1(y) \end{bmatrix}.$$

Remark 6: A tangential joint of two adjacent Coons patches with a common boundary curve is achieved when l_0, l_1 and m_0, m_1 respectively are replaced by the cubic Hermite polynomials H_0^0 and H_1^0 (cf. **18.9**) provided the longitudinal boundary curves meet tangentially and all boundary curves are differentiable. The construction of such composite surfaces was the starting point of Coons' considerations.

Remark 7: The customary two-dimensional interpolation schemes in **18.10** interpolate only a finite number of support points. For this reason they are called **finite**. Compared with them, a Coons patch and its generalization by Gordon interpolates the curves of an entire net, point by point. For this reason they are called **transfinite**.

18.12 Notes and Exercises

1. The recurrence relation of the divided differences $f_{i,\dots,k}$ can be derived from Aitken's Lemma.
2. A function $f(x)$ is a polynomial of degree at most n if the $(n+1)$th divided difference $f_{0,\dots,n}$ vanishes for every set of $n+1$ distinct support values $x_0, \dots, x_n$.

3. Let $f(x) = g(x) \cdot h(x)$. Prove the Leibniz formula

$$f_{0,\dots,n} = \sum_{i=0}^{n} g_{0,\dots,i}\, h_{i,\dots,n}$$

by induction on n.

4. Consider a curve $z(x) = \sum_i c_i f_i(x)$, where the $f_i(x)$ are some basis functions. Let the c_i depend on y, i.e., let $c_i(y) = \sum_k c_{i,k}\, g_k(y)$, where the $g_k(y)$ are some basis functions, too. Then the curve $z(x)$ will sweep out a surface,

$$z(x, y) = \sum_i \sum_k c_{i,k}\, f_i(x) g_k(y)\,.$$

Such a surface is called a **tensor product surface**.

5. Examples of such a tensor product surface are the Lagrange surface **18.10**, the Bézier surface **20.4** and the spline surface **27.3**.

19 Discrete Approximation

An **approximation** problem consists of finding a function $a(x)$ which lies close to a given function $f(x)$ in some interval and is, some respect, simpler. $a(x)$ is a **discrete approximation** if only a finite number of data about f is used to determine it. Interpolation is an example of discrete approximation. The easiest — but not always the best — approximators are polynomials.

19.1 The Taylor Polynomials

Suppose $f(x)$ is n-times continuously differentiable. Then the **nth degree Taylor polynomial** for f about some x_0 is defined as

$$\text{app}(x) := f(x_0) + \frac{1}{1!} f^{(1)}(x_0)(x - x_0)^1 + \cdots + \frac{1}{n!} f_0^{(n)}(x_0)(x - x_0)^n\,.$$

It agrees with f and the derivatives $f^{(1)}, \dots, f^{(n)}$ at x_0. For the deviation

$$R_{n+1}(x) := f(x) - \text{app}(x)$$

there is the well-known identity

$$R_{n+1}(x) = \frac{1}{(n+1)!} f^{(n+1)}(y)(x - x_0)^{n+1}$$

with $y = y(x) \in [x_0, x]$ provided that $f^{(n+1)}$ exists continuously.

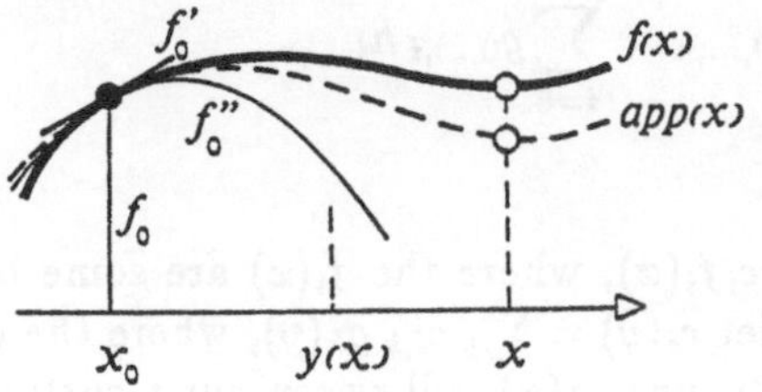

Figure 19.1
Taylor approximation

If $f^{(n+1)}(y)$ is bounded, then the error is, in principle, characterized by $(x - x_0)^{n+1}$.

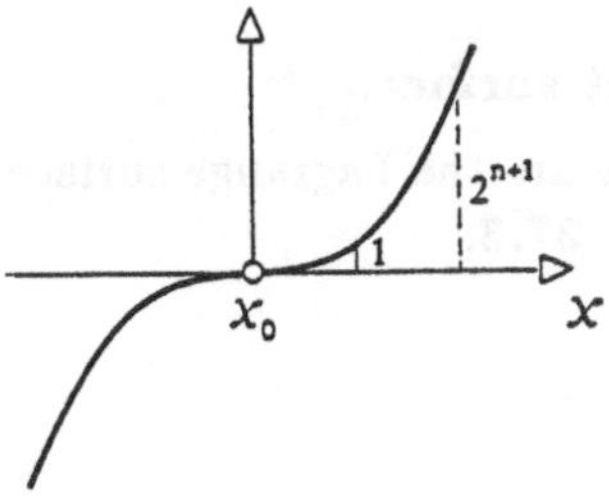

Figure 19.2
Error function (n even)

A doubling of $x - x_0$ results in a multiplication of R_{n+1} by roughly 2^{n+1}. Therefore the Taylor polynomial is useful merely in a small neighborhood of x_0. But there it is excellent.

Example 1: Consider the problem of approximating $f(x) = \sin x$ in the interval $[a, b] = [-\frac{\pi}{2}, \frac{\pi}{2}]$ by a polynomial of degree $n = 4$. The Taylor polynomial of order 4 for f about $x_0 = 0$ is

$$\text{app}(x) = x - 0.167x^3,$$

and

$$|f^{(5)}(x)| \leq 1 \text{ and } \left(\frac{\pi}{2}\right)^5 \leq 9.6$$

implies

$$|R_5| \leq \frac{1}{120} \cdot 1 \cdot 9.6 = 0.08.$$

19.2 The Interpolation Polynomial

The interpolation polynomial pol(x) which agrees with $f(x)$ at $n+1$ distinct numbers x_i in an interval $I := [a, b]$ yields often a better approximation over the entire interval than the Taylor polynomial.

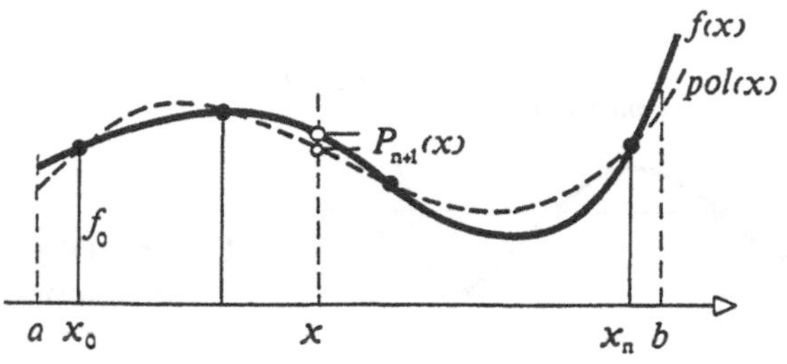

Figure 19.3
Interpolation polynomial

The deviation $P_{n+1}(x) := f(x) - \text{pol}(x)$ is considered in the next theorem.

Theorem 1: *Let $f(x)$ be $n+1$ times continuously differentiable in $I := [a, b]$ and suppose the interpolation abscissae lie in I. Then there exists a $y(x) \in I$ for every $x \in I$ such that*

$$P_{n+1}(x) = \frac{1}{(n+1)!} f^{(n+1)}(y) \cdot p_{n+1}(x)$$

where $p_{n+1}(x) := (x - x_0)(x - x_1) \cdots (x - x_n)$.

The proof is carried out in three steps:

1. $P_{n+1}(x)$ has the roots $x_0, \ldots, x_n$ due to the construction of pol(x).

2. $p_{n+1}(x) \neq 0$ for $x \neq x_i$, and there exists a γ depending on x such that

$$P_{n+1}(x) = \gamma(x) \cdot p_{n+1}(x).$$

3. Therefore

$$F(t) := P_{n+1}(t) - \gamma(x) \cdot p_{n+1}(t)$$

has $n+2$ zeros in I, namely $x_0, \ldots, x_n$ and x. As a consequence of Rolle's Theorem, then, $F'(t)$ has $n+1$ zeros in I and finally $F^{(n+1)}(t)$ one zero $y = y(x)$ in I. Differentiating F with respect to t and using $P_{n+1}(t) = f(t) - \text{pol}(t)$ gives

$$F^{(n+1)}(t) = f^{(n+1)}(t) - 0 - \gamma(x) \cdot (n+1)!$$

whence for $t = y$

$$\gamma(x) = \frac{1}{(n+1)!} f^{(n+1)}(y).$$

Again, the error is principally characterized by $p_{n+1}(x)$.

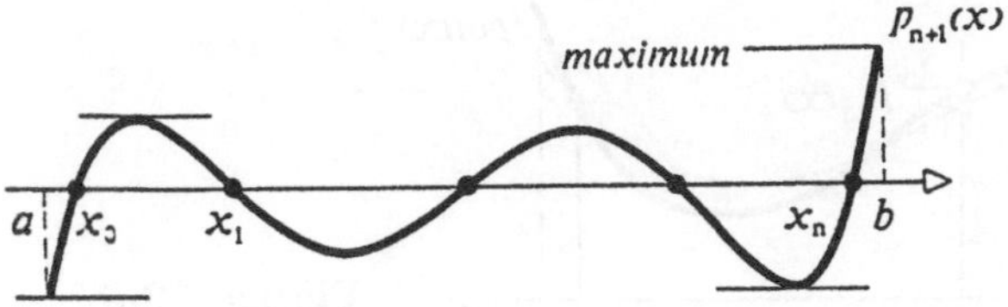

Figure 19.4
Error function

Example 2: For Example 1 with the equidistant interpolation abscissae

$$x_0 = -\frac{\pi}{2}, \quad x_1 = -\frac{\pi}{4}, \quad x_2 = 0, \quad x_3 = +\frac{\pi}{4}, \quad x_4 = +\frac{\pi}{2}$$

one obtains

$$\text{pol}(x) = 0.9882x - 0.1425x^3$$

and, because $|p_{n+1}(x)| \leq 1.0853$ in I,

$$|P_5| \leq \frac{1}{120} 1.0853 = 0.009\,.$$

Compared with the Taylor polynomial in Example 1, one decimal place is gained.

19.3 Chebyshev Approximation

Clearly, the approximation error depends very much on the support abscissae x_i. Therefore, one should choose the x_i so as to minimize the **supremum norm**

$$\|p_{n+1}(x)\| := \max_{x \in I} |p_{n+1}(x)|$$

of $p_{n+1}(x)$ in the interval I. The solution to this minimization problem is provided in the sequel. It is based on the following fact, here stated for $p_n(x)$.

Theorem 2: *The polynomial $p_n(x)$ of degree n whose zeros x_i lie on the projections of the "odd" vertices of a regular $4n$-gon onto the diameter $[a, b]$ attains absolutely equal extremal values at the $n+1$ projections y_i of the "even" vertices.*

The proof is done in **19.4**.

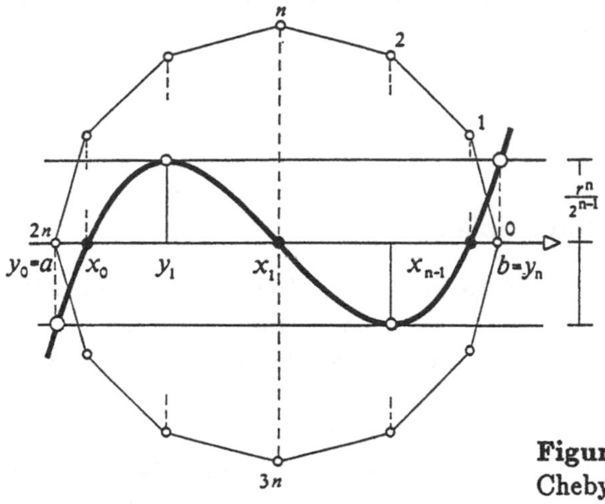

Figure 19.5
Chebyshev abscissae

Remark 1: Using the abbreviations

$$m := \frac{1}{2}(a+b) \text{ and } r := \frac{1}{2}(b-a)$$

one finds the zeros of $p_n(x)$ at

$$x_i := m - r\cos\frac{2i+1}{n}\frac{\pi}{2}, \quad i = 0, 1, \ldots, n-1,$$

and the extremal values in I at

$$y_j := m - r\cos\frac{2j}{n}\frac{\pi}{2}, \quad j = 0, 1, \ldots, n,$$

and if $p_n(x)$ is a monic polynomial

$$\|p_n(x)\| = 2\left(\frac{r}{2}\right)^n, \quad x \in I,$$

see **19.5**. The zeros x_i of $p_n(x)$ are called **Chebyshev points.**

Example 3: Consider once more Example 1. The interpolation polynomial which interpolates $f(x) = \sin x$ at the Chebyshev points

$$x_0 = -0.951\frac{\pi}{2}, \quad x_1 = -0.588\frac{\pi}{2}, \quad x_2 = 0, \quad x_3 = 0.588\frac{\pi}{2}, \quad x_4 = 0.951\frac{\pi}{2}$$

in the interval $I = \left[-\frac{\pi}{2}, \frac{\pi}{2}\right]$ is

$$\text{pol}(x) = 0.9855x - 0.1424x^3.$$

Since $|p_5(x)| \le 0.5977$ in I, the maximal deviation is halved as compared to Example 2,

$$|P_5| \le \frac{1}{120} 0.5977 = 0.005\,.$$

19.4 Chebyshev Polynomials

It suffices to prove Theorem 2 for the interval $[-1,1]$ since the result can be generalized to any other interval by just a linear transformation. By a well-known formula from trigonometry, $\cos n\varphi$ is expressible as a polynomial in $x = \cos\varphi$ of degree n,

$$\cos n\varphi = T_n(\cos\varphi)\,.$$

The polynomials $T_n(x)$ are called **Chebyshev polynomials**. They vanish when $\cos n\varphi = 0$, i.e., if $\varphi = -(2i+1)\pi/2n$. So, the zeros of $T_n(x)$ are

$$x_i = -\cos\frac{2i+1}{n}\frac{\pi}{2}, \quad i = 0, 1, \ldots, n-1\,.$$

Similarly, $T_n(x)$ is found to attain the maxima and minima ± 1 at

$$y_j = -\cos\frac{2j}{n}\frac{\pi}{2}, \quad j = 0, 1, \ldots, n\,.$$

This, basically, establishes the proof of Theorem 2.

The Chebyshev polynomials can be obtained recursively by virtue of the trigonometric identity

(1) $$\cos(\alpha-\beta) + \cos(\alpha+\beta) = 2\cos\alpha\cos\beta\,.$$

For $\alpha = n\varphi$ and $\beta = \varphi$, this identity is translated into

$$T_{n-1}(x) + T_{n+1}(x) = 2xT_n(x)\,.$$

This recurrence relation together with

$$T_0(x) := \cos 0 = 1\,,$$
$$T_1(x) := \cos\varphi = x$$

provides a tool to construct all Chebyshev polynomials, namely one gets

(2) $$\begin{bmatrix} T_0(x) \\ T_1(x) \\ T_2(x) \\ T_3(x) \\ T_4(x) \\ T_5(x) \\ \vdots \end{bmatrix} = \begin{bmatrix} \cos 0 \\ \cos\varphi \\ \cos 2\varphi \\ \cos 3\varphi \\ \cos 4\varphi \\ \cos 5\varphi \\ \vdots \end{bmatrix} = \begin{bmatrix} 1 & & & & & & \\ 0 & 1 & & & & & \\ -1 & 0 & 2 & & & & \\ 0 & -3 & 0 & 4 & & & \\ 1 & 0 & -8 & 0 & 8 & & \\ 0 & 5 & 0 & -20 & 0 & 16 & \\ \vdots & \vdots & \vdots & \vdots & & & \ddots \end{bmatrix} \begin{bmatrix} 1 \\ x \\ x^2 \\ x^3 \\ x^4 \\ x^5 \\ \vdots \end{bmatrix}.$$

Obviously, 2^{n-1} forms the leading coefficient of $T_n(x)$, $n \geq 1$.

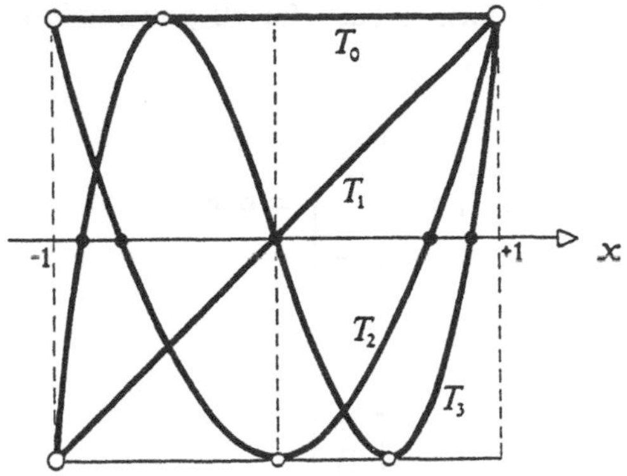

Figure 19.6
Chebyshev polynomials

19.5 The Minimum Property

The Chebyshev polynomials have the following remarkable property:

Theorem 3: $2^{1-n}T_n(x)$ *has the least supremum norm in* $[-1,1]$ *among all monic polynomials of degree* $\leq n$.

The proof is indirect. If the monic polynomial pol(x) had a supremum norm on $[-1,1]$ less than that of $2^{1-n}T_n$, then the difference $2^{1-n}T_n-$pol would have to have a zero between any two of the $n+1$ values y_i at which T_n assumes its alternating extrema. Whence one concludes that $2^{1-n}T_n(x) -$ pol(x) changes signs n-times in $[-1,1]$. But this is impossible since the degree of the polynomial $2^{1-n}T_n(x)-$pol(x) is $n-1$ at most.

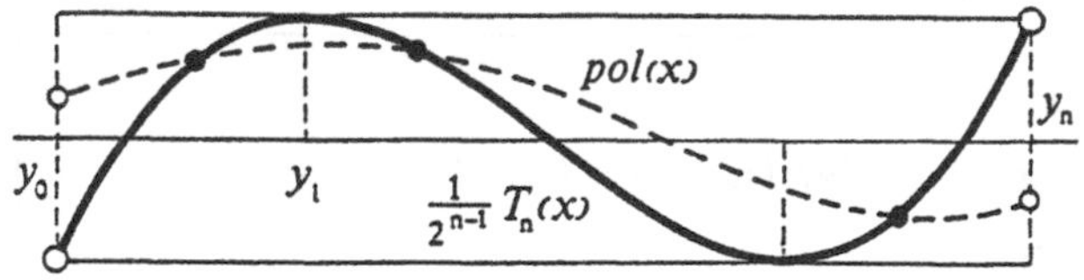

Figure 19.7
Minimum property

19.6 Expanding by Chebyshev Polynomials

The Chebyshev polynomials $T_0, \dots, T_n$ form a basis for all polynomials of degree $\le n$, i.e., every polynomial

$$\text{pol}(x) = [c_0, c_1, \dots, c_n] \begin{bmatrix} x^0 \\ x^1 \\ \vdots \\ x^n \end{bmatrix} = \boldsymbol{c}^t \boldsymbol{x}$$

has a unique representation

$$\text{pol}(x) = [a_0, a_1, \dots, a_n] \begin{bmatrix} T_0 \\ T_1 \\ \vdots \\ T_n \end{bmatrix} = \boldsymbol{a}^t \boldsymbol{t}\,.$$

The coefficients $a_0, \dots, a_n$ can be found with the aid of relation (2). Let L denote the $(n+1) \times (n+1)$-submatrix of the linear system (2) given by the first $n+1$ rows and columns.

Then one has

$$\boldsymbol{a}^t \boldsymbol{t} = \boldsymbol{a}^t L \boldsymbol{x}\,,$$

and a comparison with $\boldsymbol{c}^t \boldsymbol{x}$ shows that

$$L^t \boldsymbol{a} = \boldsymbol{c}\,.$$

This equation can be solved by backward substitution. On transcribing the algorithm in **3.1** with the notation used here one gets:

Expanding by Chebyshev Polynomials

Input:	$\boldsymbol{c} = [c_0, \dots, c_n]^t$,
	$L = [l_{ik}]$ lower triangular $(n+1) \times (n+1)$-matrix
	with $\boldsymbol{t}(x) = L\boldsymbol{x}$
Output:	$\boldsymbol{a} = [a_0, \dots, a_n]^t$ such that $\boldsymbol{a}^t \boldsymbol{t}(x) = \boldsymbol{c}^t \boldsymbol{x}$

1	For $k = n, n-1, \dots, 1$
2	$\quad a_k := \dfrac{1}{l_{k,k}}(c_k - l_{k+1,k}a_{k+1} - \cdots - l_{n,k}a_n)\,.$

19.7 Economization of Polynomials

In the evaluation of a polynomial a low degree is certainly preferable to a high degree. Therefore, it can be desirable to replace a polynomial of degree n by a polynomial of lower degree, of course without having too much of a deviation. For this task the Chebyshev expansion proves very useful. One readily verifies that the best one can do in reducing the degree by one is to simply drop the last term $a_n T_n$. The error $|a_n|$ is minimal over the interval $[-1, 1]$.

This process can be continued as long as the cumulative error

$$|a_n| + |a_{n-1}| + \cdots + |a_k|$$

does not exceed some given value $\varepsilon > 0$. The reduced polynomial $a_0 T_0 + \cdots + a_k T_k$ is generated by the algorithm **Expanding by Chebyshev Polynomials** with the additional instruction

3	If $\sum_{i=k}^{n} \lvert a_i \rvert > \varepsilon$, save k and **stop**.

The output is $\boldsymbol{a} = [a_0, \ldots, a_k]$.

Example 4: Consider $f(x) = \sin \frac{\pi}{2} x$ in $[-1, 1]$. The 6th degree Taylor polynomial around $x_0 = 0$ is

$$\begin{aligned} \text{app}(x) &= 1.571x - 0.646x^3 + 0.080x^5 \\ &= 1.137 \cdot T_1 - 0.137 \cdot T_3 + 0.005 \cdot T_5 . \end{aligned}$$

Since $|R_7| \leq \frac{1}{5040}(\frac{\pi}{2})^7 = 0.005$ and $a_6 = 0$, $a_5 = 0.005$, the error made in approximating $\sin \frac{\pi}{2} x$ by

$$\begin{aligned} \text{pol}(x) &= 1.137 \cdot T_1 - 0.137 \cdot T_3 \\ &= 1.546x - 0.546x^3 \end{aligned}$$

is at most $|R_7| + |a_6| + |a_5| \leq 0.01$. For a comparison with Examples 1, 2, and 3, note that the interval here has been transformed into $[-1, 1]$.

19.8 Least Squares Method

The interpolation polynomial

$$\text{pol}(x) = c_0 + c_1 x + \cdots + c_n x^n$$

which approximates a function $f(x)$ is uniquely defined by $n+1$ interpolation points. However, one may use more than $n+1$ values of f to construct pol(x), say $f(x_i)$, $i = 0, \ldots, m$, with $m > n$.

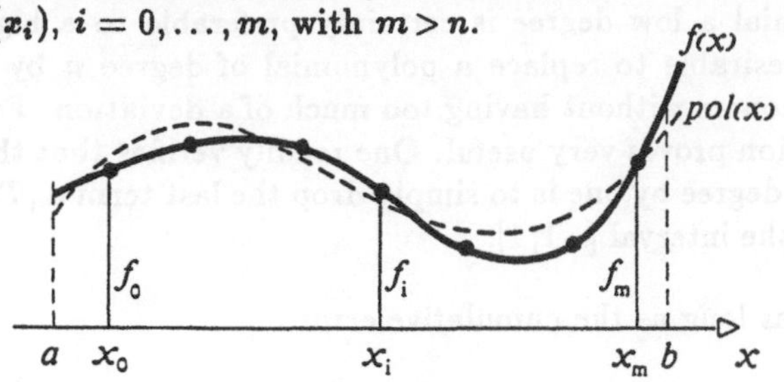

Figure 19.8
Approximation polynomial

Of course, in most cases, pol(x) will no longer interpolate all $m+1$ function values. In other words, the system of linear equations pol(x_i) $= f(x_i)$, i.e.,

$$\begin{bmatrix} 1 & x_0 & \cdots & x_0^n \\ \vdots & \vdots & & \vdots \\ 1 & x_m & \cdots & x_m^n \end{bmatrix} \begin{bmatrix} c_0 \\ \vdots \\ c_n \end{bmatrix} = \begin{bmatrix} f_0 \\ \vdots \\ f_m \end{bmatrix},$$

for the coefficients c_k is, in general, overdetermined.

This system can be solved by the methods of **9.2**. The solution thus obtained is called a **least squares fit**. Since the normal equations are usually ill-conditioned, an orthogonalization is recommended, it corresponds to a basis transformation for the representation of pol(x). The new basis depends on the points x_i.

19.9 The Orthogonality of Chebyshev Polynomials

In particular, let x_i, $i = 0, 1, \ldots, m$, be the zeros of the Chebyshev polynomial $T_{m+1}(x)$. Then one has:

Theorem 4: *The square matrix formed by the $T_k(x_i)$ is orthogonal, namely its columns t_k satisfy*

$$t_r^t t_s = \begin{cases} m+1 & \text{for } r = s = 0\,, \\ \frac{1}{2}(m+1) & \text{for } r = s \neq 0\,, \\ 0 & \text{for } r \neq s\,. \end{cases}$$

For the proof one takes advantage of the representation $T_k(x_i) = \cos k\varphi_i$ with $x_i = \cos\varphi_i$ and employs the identity (1) in **19.4**. This leads to

$$\begin{aligned} t_r^t t_s &= \sum_{i=0}^{m} \cos r\varphi_i \cos s\varphi_i \\ &= \frac{1}{2}\sum_{i=0}^{m} \cos(r+s)\varphi_i + \frac{1}{2}\sum_{i=0}^{m} \cos(r-s)\varphi_i\,. \end{aligned}$$

All what remains to be done is to apply and to establish

$$(3) \qquad \sum_{i=0}^{m} \cos(r \pm s)\varphi_i = \begin{cases} m+1 & \text{for } r \pm s = 0 \\ 0 & \text{for } r \pm s \neq 0 \,. \end{cases}$$

From Theorem 2 follows that $(\cos\varphi_i, \pm\sin\varphi_i)$, $i = 0, \ldots, m$, are the vertices of a regular $2(m+1)$-gon. More general, also for all integers $\mu \neq 0$, $(\cos\mu\varphi_i, \pm\sin\mu\varphi_i)$ are the vertices of a regular $2(m+1)$-gon whose center is the origin. It may be folded and have multiple vertices, though. Now, (3) ensues from the intuitively obvious theorem that the center of a regular n-gon equals the average of its vertices. This completes the proof.

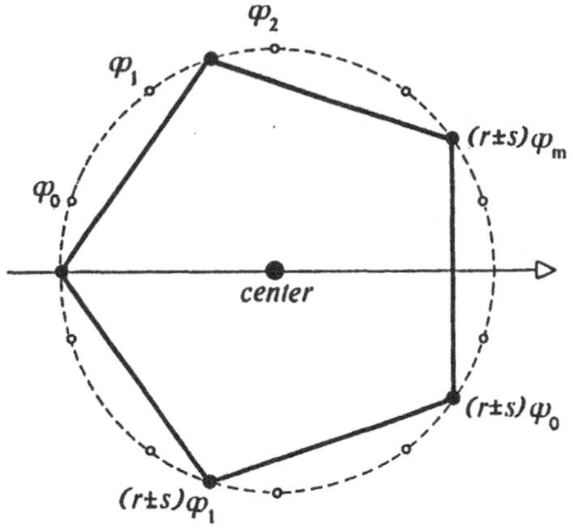

Figure 19.9
Center of a regular $2(m+1)$-gon

Example 5: Let $m = 4$ and consider the problem of finding the cubic polynomial

$$\text{pol}(x) = a_0T_0 + a_1T_1(x) + a_2T_2(x) + a_3T_3(x)$$

which approximates $\sin\frac{\pi}{2}x$ in $[-1, 1]$ while minimizing the sum of the squared deviations at the zeros of $T_5(x)$,

$$x_0 = -0.851\,, \quad x_1 = -0.588\,, \quad x_2 = 0\,, \quad x_3 = 0.588\,, \quad x_4 = 0.851\,.$$

Because of Theorem 4, the normal equations are

$$\begin{bmatrix} 5 & & & \\ & 2.5 & & \\ & & 2.5 & \\ & & & 2.5 \end{bmatrix} \begin{bmatrix} a_0 \\ a_1 \\ a_2 \\ a_3 \end{bmatrix} = \begin{bmatrix} 1 & 1 & 1 & 1 & 1 \\ -0.98 & -0.59 & 0 & 0.59 & 0.85 \\ 0.81 & -0.31 & -1 & -0.31 & 0.81 \\ -0.59 & 0.85 & 0 & -0.95 & 0.59 \end{bmatrix} \begin{bmatrix} -0.997 \\ -0.798 \\ 0 \\ 0.798 \\ 0.997 \end{bmatrix} .$$

Hence, the solution is found to be

$$\begin{aligned} \text{pol}(x) &= 1.134 \cdot T_1 - 0.138 \cdot T_3 \\ &= 1.548x - 0.552x^3 \,. \end{aligned}$$

The maximum error occurs at approximately $x = \pm 0.3$ and is smaller than 0.005. Here, in this example, the sum of the squared deviations even vanishes (cf. Example 3 after the necessary interval transformation).

19.10 Notes and Exercises

1. **Verify the formula for the remainder of the Taylor expansion.**
2. **Analogously to Horner's scheme, $\mathrm{pol}(x) = a_0T_0(x) + \cdots + a_nT_n(x)$ can be evaluated by the algorithm of Clenshaw:**

Clenshaw

```
Input:   a_0, a_1, ..., a_n; x, n >= 2
Output:  p = a_0 T_0(x) + ... + a_n T_n(x)

Set b_n := a_n, b_{n+1} := 0 .
For k = n-1, n-2, ..., 1
    b_k := -b_{k+2} + 2x b_{k+1} + a_k .
Set p := -b_2 + x b_1 + a_0 .
```

3. **To perform the algorithm of Clenshaw manually it is convenient to use the following scheme**

-1	$2x$	a_n	a_{n-1}	$\cdots$	a_2	a_1	a_0
0	0	b_n	b_{n-1}	$\cdots$	b_2	b_1	p

with the rules

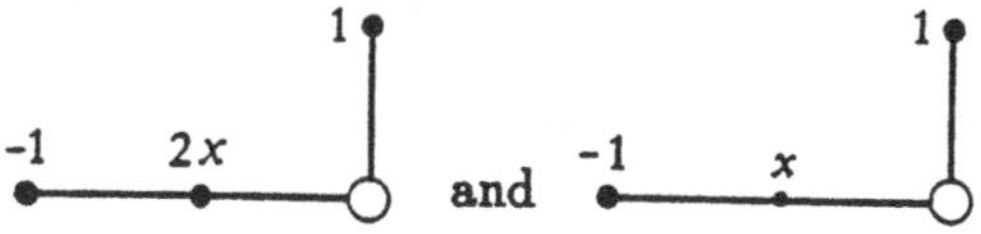

20 Polynomials in Bézier Form

Lagrange or Chebyshev polynomials are not well suited for the purpose of geometric modeling with polynomial or rational curves. One needs basis polynomials which do not oscillate in $I = [a, b]$ and have only one maximum there. **Bernstein polynomials** are such polynomials.

20.1 Bernstein Polynomials

The summands of the binomial expansion

$$1 = ((1-\lambda) + \lambda)^n = \sum_{r=0}^{n} \binom{n}{r} (1-\lambda)^{n-r} \lambda^r$$

are polynomials of degree n in λ. These polynomials

$$B_r^n(\lambda) := \binom{n}{r} (1-\lambda)^{n-r} \lambda^r, \quad r = 0, \ldots, n,$$

are called **Bernstein polynomials.**

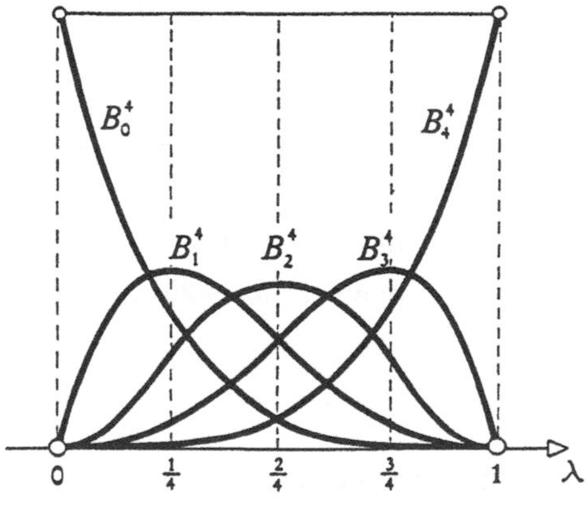

Figure 20.1
Bernstein polynomials of degree 4

The Bernstein polynomials have the following properties which can easily be verified:

1. $B_r^n(\lambda)$ has an r-fold root at $\lambda = 0$.
2. $B_r^n(\lambda)$ has an $(n-r)$-fold root at $\lambda = 1$.
3. $B_r^n(\lambda)$ has only one maximum in $I := [0, 1]$, namely at $\lambda = \frac{r}{n}$.

For this reason one works with the Bernstein polynomials solely in the interval $I := [0, 1]$.

20.2 Polynomials in Bézier Form

The Bernstein polynomials of degree n are linearly independent because of the properties 1 and 2. Therefore they form a basis for all polynomials of degree at most n. The representation

$$\mathrm{pol}(\lambda) = b_0 B_0^n(\lambda) + \cdots + b_n B_n^n(\lambda)$$

is called **Bézier representation**, the coefficients b_i **Bézier points** and the polygon with the vertices $(\frac{i}{n}, b_i)$ **Bézier polygon.**

The derivative of a polynomial in its Bézier representation is obtained by differentiating the Bernstein polynomials. One has

$$\mathrm{pol}'(\lambda) = n(b_1 - b_0)B_0^{n-1}(\lambda) + \cdots + n(b_n - b_{n-1})B_{n-1}^{n-1}(\lambda)$$

and in general for the kth derivative

$$\mathrm{pol}^{(k)}(\lambda) = \frac{n!}{(n-k)!} \sum_{i=0}^{n-k} \Delta^k b_i \cdot B_i^{n-k}(\lambda),$$

where Δ^k denotes the kth forward difference, **13.4**. This fact together with the properties 1 and 2 of the Bernstein polynomials constitutes the following theorem.

Theorem 1: *The kth derivative of a polynomial at the endpoints $\lambda = 0$ and $\lambda = 1$ of I depends merely on its Bézier points $b_0, \dots, b_k$ and $b_{n-k}, \dots, b_n$ respectively.*

In particular

$$\begin{aligned} \mathrm{pol}(0) &= b_0, & \mathrm{pol}(1) &= b_n, \\ \mathrm{pol}'(0) &= n(b_1 - b_0), & \mathrm{pol}'(1) &= n(b_n - b_{n-1}), \\ \mathrm{pol}''(0) &= n(n-1)(b_2 - 2b_1 + b_0), & \mathrm{pol}''(1) &= n(n-1)(b_n - 2b_{n-1} + b_{n-2}). \end{aligned}$$

20.3 The Construction of Position and Tangent

1959, de Casteljau developed a construction of the values and derivatives of a polynomial at any λ which is an **iterated linear interpolation** of its Bézier points.

Theorem 2: *The polynomials $b_{r,\dots,s}(\lambda) := \sum_{i=r}^{s} b_i B_{i-r}^{s-r}(\lambda)$ of degree $s - r$ which are controlled by the Bézier points $b_r, \dots, b_s$ satisfy the recurrence relation*

$$b_{r,\dots,s}(\lambda) = (1-\lambda) b_{r,\dots,s-1}(\lambda) + \lambda b_{r+1,\dots,s}(\lambda).$$

The proof can be based on the identities

$$(1-\lambda)B_0^n = B_0^{n+1}\,, \qquad (1-\lambda)B_r^n + \lambda B_{r-1}^n = B_r^{n+1}\,, \qquad \lambda B_n^n = B_{n+1}^{n+1}$$

which themselves can easily be established. In particular, one gets $\mathrm{pol}(\lambda) = b_{0,\ldots,n}$.

Furthermore, with the notations $\Delta b_{r,\ldots,s} := b_{r+1,\ldots,s} - b_{r,\ldots,s-1}$, etc., one has

Theorem 3: *The derivatives of* $\mathrm{pol}(\lambda)$ *are easily obtained from its Bézier points*

$$\mathrm{pol}^{(k)}(\lambda) = \frac{n!}{(n-k)!}\Delta^k b_{0,\ldots,n-k}(\lambda)\,.$$

For the proof, one observes that the forward differences are of the form $\Delta^k b_i = \sum_{j=0}^{k} \delta_j^k\, b_{i+j}$ with certain integers δ_j^k. Hence,

$$\begin{aligned}\mathrm{pol}^{(k)}(\lambda) &= \frac{n!}{(n-k)!}\sum_{i=0}^{n-k}\sum_{j=0}^{k}\delta_j^k\, b_{i+j}\, B_i^{n-k}\\ &= \frac{n!}{(n-k)!}\sum_{j=0}^{k}\delta_j^k\sum_{i=0}^{n-k} b_{i+j}\, B_i^{n-k}\\ &= \frac{n!}{(n-k)!}\Delta^k b_{0,\ldots,n-k}\,.\end{aligned}$$

The Neville-like **scheme of de Casteljau** helps to organize the calculation of the $b_{r,\ldots,s}$:

$$\begin{array}{ll|llllll}
 & & b_0 & & & & & \\
 & r\ - & b_1 & b_{0,1} & & & & \\
1-\lambda & & b_2 & \underline{b_{1,2}} & b_{0,1,2} & & & \\
\lambda & s\ - & b_3 & \underline{b_{2,3}} & \underline{\underline{b_{1,2,3}}} & b_{0,1,2,3} & & \\
 & & \vdots & \vdots & & & \ddots & \\
 & & b_n & b_{n-1,n} & \cdots & & \cdots & b_{0,\ldots,n} = \mathrm{pol}(\lambda)\,.
\end{array}$$

The first column contains the Bézier points b_i. The other $b_{r,\ldots,s}$ and in particular $b_{0,\ldots,n} = \mathrm{pol}(\lambda)$ are determined row after row or column after column as follows:

de Casteljau

Input:	$b_r, \dots, b_s; \lambda$
Output:	$b_{r,\dots,s}$

For $k = r+1, r+2, \dots, s$
  and $i = k-1, k-2, \dots, r$
    $b_{i,\dots,k} := (1-\lambda) b_{i,\dots,k-1} + \lambda b_{i+1,\dots,k}$.

Remark 1: For the construction on a drawing board or on a screen one starts out from the Bézier polygon, i.e., from the points $(\frac{i}{n}, b_i)$ instead of the coefficients b_i. One can easily verify then: The construction of de Casteljau yields the point $(\lambda, \text{pol}(\lambda))$ because of the identity

$$\lambda = \sum_{i=0}^{n} \frac{i}{n} B_i^n(\lambda) .$$

In particular, $(0, b_0)$ lies on the curve, the line through $(0, b_0)$ and $(\frac{1}{n}, b_1)$ is the tangent at $(0, b_0)$, etc. In general, the Bézier polygon gives an impression of the approximate shape of the corresponding polynomial.

Example 1: Consider the cubic polynomial

$$\text{pol}(\lambda) = 1 \cdot (1-\lambda)^3 + 4 \cdot 3(1-\lambda)^2\lambda + 3 \cdot 3(1-\lambda)\lambda^2 + 0 \cdot \lambda^3$$

with the Bézier points $b_0 = 1$, $b_1 = 4$, $b_2 = 3$, $b_3 = 0$. For $\lambda = 0.4$ the scheme of de Casteljau is the following:

	1			
0.6	4	2.2		
0.4	3	3.6	2.76	
	0	1.8	2.88	2.808 .

Hence, $\text{pol}(0.4) = 2.808$ and $\text{pol}'(0.4) = 3(2.88 - 2.76) = 0.36$. In the figure $b_{i,i+1,\dots,j}$ is abbreviated by $b_{i,j}$.

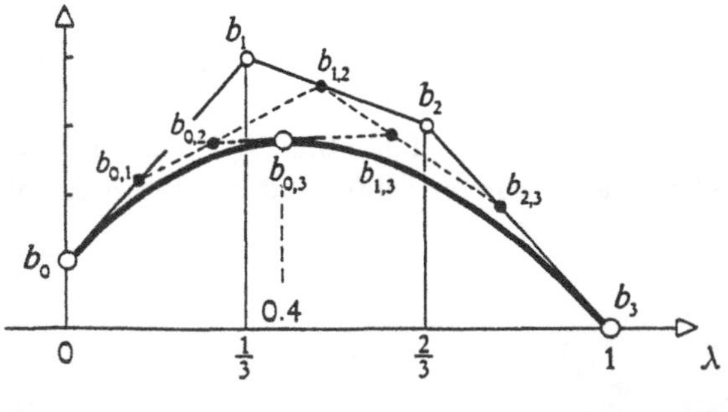

Figure 20.2
Cubic Bézier polynomial

Example 2: If the Bézier points are 0, 1, 0, 0, one obtains the Bernstein polynomial $\mathrm{pol}(\lambda) = B_1^3(\lambda)$. The construction of de Casteljau for $\lambda = \frac{1}{3}$ yields the maximum $B_1^3(\frac{1}{3}) = \frac{4}{9}$.

20.4 Bézier Surfaces

The products of the Bernstein polynomials $B_r^n(\lambda)$ and $B_s^m(\mu)$ form a basis of the polynomials of degree at most n and m in the variables λ and μ

$$\mathrm{pol}(\lambda, \mu) = \sum_{i=0}^{n}\sum_{k=0}^{m} b_{i,k}\, B_k^m(\mu) B_i^n(\lambda)\,.$$

Again, the coefficients are called **Bézier points.** They form the **Bézier matrix** B. The **Bézier representation of polynomials** in two variables has the same pleasant properties as for polynomials in one variable. Especially, one can derive immediately from the representation above the following:

Theorem 4: *The univariate polynomials* $\mathrm{pol}(\lambda, \mu)$, $\mu =$ *constant, have the Bézier points*

$$b_i(\mu) = \sum_{k=0}^{m} b_{i,k}\, B_k^m(\mu)\,.$$

Thus, de Casteljau's construction for $\mu =$ constant applied to each row i of the matrix B which can be seen as a de Casteljau contruction on the $m+1$ columns of the matrix B generates the Bézier points of $\mathrm{pol}(\lambda, \mu)$, $\mu =$ constant.

For a graphical display one associates the Bézier points $b_{i,k}$ with the abscissae $(\frac{i}{n}, \frac{k}{m})$, etc.

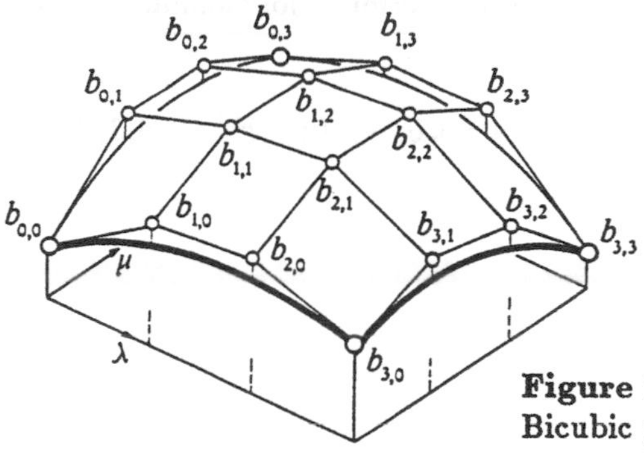

Figure 20.3
Bicubic polynomial with its Bézier points

20.5 Notes and Exercises

1. It holds

$$1 = \sum_{i=0}^{n} B_i^n(\lambda), \quad \lambda = \sum_{i=0}^{n} \frac{i}{n} B_i^n(\lambda), \quad \lambda^2 = \sum_{i=0}^{n} \frac{i(i-1)}{n(n-1)} B_i^n(\lambda),$$

   etc.

2. Bernstein's proof of **Weierstrass' Approximation Theorem** uses the following property. The sequence of polynomials of degree n

$$B^n f := \sum_{r=0}^{n} f\left(\frac{r}{n}\right) B_r^n$$

   converges uniformly to f in $[0,1]$ as $n \to \infty$ for all functions f which are continuous in $[0,1]$.

3. One has $\mathrm{pol}(\lambda) = \sum_{r=0}^{n-k} b_{r,\ldots,r+k}(\lambda)\, B_r^{n-k}$, $k = 0, \ldots, n$, and therefore

$$\mathrm{pol}(\lambda) = \sum_{r=0}^{n-k} \sum_{i=r}^{r+k} b_i\, B_{i-r}^k(\lambda) B_r^{n-k}(\lambda).$$

4. If one replaces b_i by columns of coordinates, then $\mathrm{pol}(\lambda)$ presents a **Bézier curve**. This is the origin of the term Bézier point for b_i.

5. The construction of de Casteljau generates the column $\mathbf{pol}(\lambda)$ from the columns $\mathbf{b}_i$.

6. If l denotes the edge opposite $\mathbf{b}_2$ of the triangle $\mathbf{b}_0, \mathbf{b}_1, \mathbf{b}_2$ and h the corresponding height, then the radius of curvature at $\mathbf{b}_0$ of $\mathbf{pol}(\lambda)$ equals $R = \frac{n}{n-1} \frac{l^2}{h}$.

7. The Bézier points b_0, b_1, b_2, b_3 of the cubic Hermite polynomials H_i^j in **18.9** with respect to the interval $[0,1]$ are

$$H_0^0 : 1, 1, 0, 0; \quad H_0^1 : 0, \tfrac{1}{3}, 0, 0; \quad H_1^1 : 0, 0, -\tfrac{1}{3}, 0; \quad H_1^0 : 0, 0, 1, 1.$$

21 Splines

In general, a high degree interpolation polynomial does not yield a better approximation of a function $f(x)$ than a low degree polynomial. It is better, then, to subdivide $f(x)$ and to approximate all segments by polynomials of some fixed degree which satisfy certain differentiability conditions at the knots.

21.1 Bézier Curves

The simplest way to subdivide a function over the interval $[0, m]$ is to split it at $k = 1, \ldots, m-1$. These points are called **knots**. They sever the **segment** $k-1$ from the segment k. These segments can, for instance, be represented approximately by polynomials of degree n. In order to use the Bézier representation **20.2**, one introduces for each segment $k = 0, 1, \ldots, m-1$ the **local parameter**

$$\lambda = x - k, \quad \lambda \in [0,1],$$

and denotes the Bézier points of the kth segment successively by $b_{nk}, b_{nk+1}, \ldots, b_{nk+n}$. Here, the assumption is made that the last Bézier point b_{nk+n} of the kth **Bézier segment** coincides with the first Bézier point $b_{n(k+1)}$ of the $(k+1)$th segment, $k = 0, \ldots, m-1$. The composite curve is then continuous and defined by its $nm+1$ Bézier points $b_0, \ldots, b_{nm}$. It is referred to as a **Bézier curve**. For its graphical display one can, as in **20.3**, associate the b_i with the abscissae $\frac{i}{n}$.

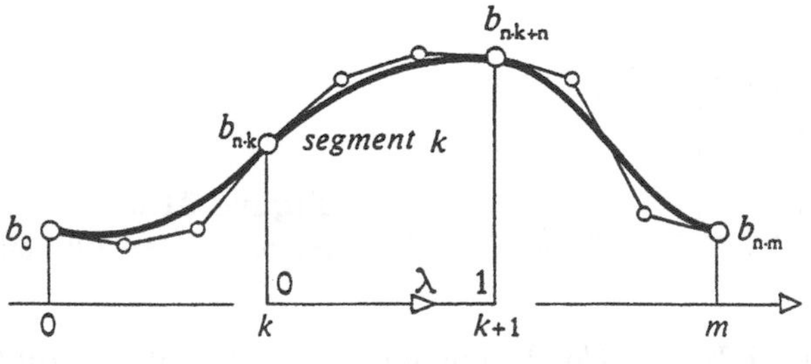

Figure 21.1
Bézier curve

Clearly, a Bézier curve can be evaluated at every $x \in [0, m]$ by means of de Casteljau's algorithm. First the interval $[k, k+1]$ containing x has to be determined. Then the algorithm **de Casteljau** has to be applied to the Bézier points $b_{nk}, \ldots, b_{nk+n}$.

21.2 Differentiability Conditions

The requirement of first or higher order differentiability of a Bézier curve (at its knots) can be described by certain conditions on successive **Bézier points**.

First order differentiability. The first derivatives at a common knot of two segments are identical according to **20.2** if $n(b_{nk} - b_{nk-1}) = n(b_{nk+1} - b_{nk})$, i.e., if

$$2b_{nk} = b_{nk-1} + b_{nk+1},$$

i.e., b_{nk} is the **midpoint** of b_{nk-1} and b_{nk+1}. The same is true for the abscissae k and $k \pm \frac{1}{n}$.

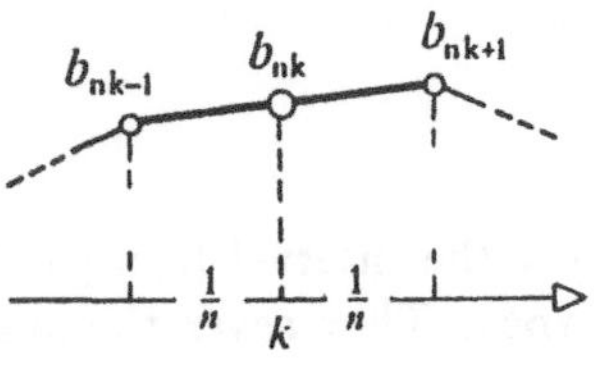

Figure 21.2
First order differentiability

Second order differentiability. The second derivatives are equal if

$$n(n-1)(b_{nk-2} - 2b_{nk-1} + b_{nk}) = n(n-1)(b_{nk} - 2b_{nk+1} + b_{nk+2}).$$

This means with the auxiliary point d_k if

$$2b_{nk-1} - b_{nk-2} = d_k = 2b_{nk+1} - b_{nk+2}.$$

$b_{nk\pm1}$ is the midpoint of d_k and $b_{nk\pm2}$. The same is true for the abscissae if the b_i are associated with the abscissae $\frac{i}{n}$ and the d_k with the abscissae k. This, together with the condition for the first order differentiability, is graphically shown in the figure.

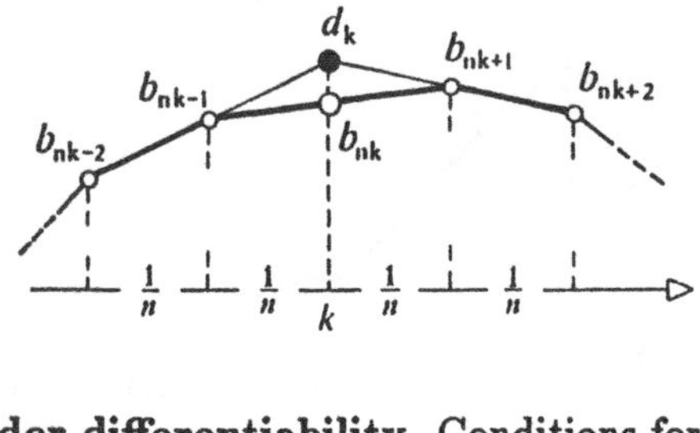

Figure 21.3
Second order differentiability

Higher order differentiability. Conditions for higher order differentiability can be formulated analogously.

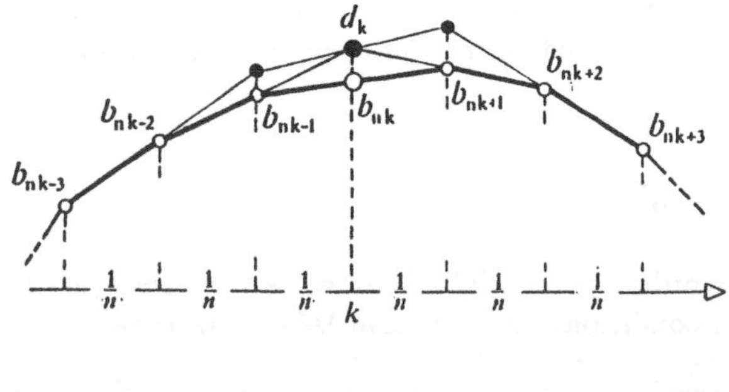

Figure 21.4
Third order differentiability

21.3 Cubic Splines

A segmented curve with polynomial segments of degree at most n is called a (polynomial) spline. There are advantages to present splines as Bézier curves. Most common are cubic segments, i.e., $n = 3$. For many applications they are both sufficiently stiff and flexible as well.

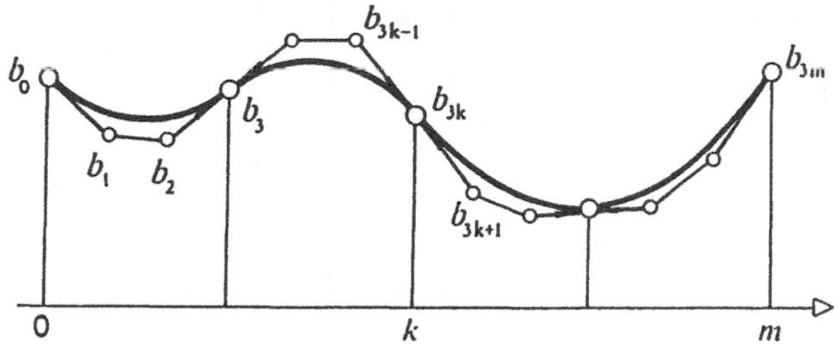

Figure 21.5 Differentiable cubic spline

Cubic C^1-splines. At all interior knots $k = 1, 2, \ldots, m-1$ one has for a differentiable cubic spline

(1) $$2b_{3k} = b_{3k-1} + b_{3k+1}\,.$$

A cubic C^1-spline is uniquely defined by its values b_{3k} and slopes at the $m+1$ knots $k = 0, \ldots, m$ according to **18.7**.

Cubic C^2-splines. Let $s(x)$ be a twice differentiable cubic spline. Using the auxiliary points d_k from **21.2**, one can derive the identities

(2) $$\begin{aligned} 3b_{3k-1} &= d_{k-1} + 2d_k \\ 3b_{3k+1} &= \qquad\quad\; 2d_k + d_{k+1}\,. \end{aligned}$$

This means that b_{3k+1} and b_{3k+2} divide the line between d_k and d_{k+1} into three equal parts. This fact allows to introduce the d_k, $k = 0, \ldots, m$, as parameters of a cubic spline.

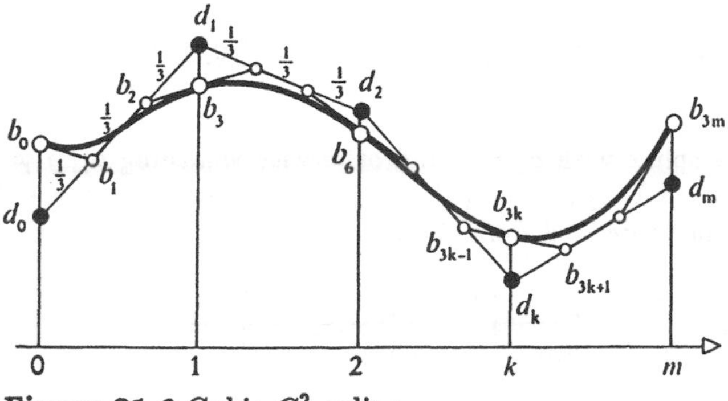

Figure 21.6 Cubic C^2-spline

Adding the equations (2) and using (1) establishes at all interior knots the identity

$$d_{k-1} + 4d_k + d_{k+1} = 6b_{3k} \,.$$

These identities together with the equations (2) for b_1 and b_{3m-1} form a tridiagonal linear system for the parameters $d_0, \ldots, d_m$:

$$\begin{bmatrix} 2 & 1 & & & \\ 1 & 4 & 1 & & \\ & & \ddots & & \\ & & 1 & 4 & 1 \\ & & & 1 & 2 \end{bmatrix} \begin{bmatrix} d_0 \\ d_1 \\ \vdots \\ d_{m-1} \\ d_m \end{bmatrix} = \begin{bmatrix} 3b_1 \\ 6b_3 \\ \vdots \\ 6b_{3m-3} \\ 3b_{3m-1} \end{bmatrix} .$$

The solution $d_0, \ldots, d_m$ determines a unique cubic C^2-spline $s(x)$ which passes through the points (k, b_{3k}), $k = 0, \ldots, m$, and has prescribed slopes at the **endpoints**. It is called a **cubic spline interpolant**. The slopes at the endpoints are given by b_1 and b_{3m-1} as mentioned in Remark 1 in **20.3**. All the other Bézier points $b_{3k\pm1}$ of $s(x)$ are found by (2).

Remark 1: If only $d_0, \ldots, d_m$ and b_0, b_{3m} are given, one determines first the inner Bézier points $b_{3k\pm1}$ of all segments by (2) and then the Bézier points b_{3k}, $k = 1, \ldots, m-1$, by (1).

Remark 2: Rather than prescribing the slopes of $s(x)$ at the endpoints, one can require $d_0 = b_0$ and $d_m = b_{3m}$. The spline interpolant $s(x)$ is then called **natural**. A cubic natural spline satisfies

$$s''(0) = s''(m) = 0 \,.$$

Remark 3: A cubic C^2-spline $s(x)$ over $\mathbb{R}$ with $d_{m+i} = d_i$ and $b_{3m+i} = b_i$, $i = \ldots, 0, 1, 2, \ldots$ is known as a **periodic spline**. It has the property

$$s(0) = s(m) \,, \quad s'(0) = s'(m) \text{ and } s''(0) = s''(m) \,.$$

Example 1: The spline with $d_k = 1$ and otherwise vanishing d_j, $j \neq k$, is a basis spline called the kth **B-spline** $N_k(x)$. The Bézier points of a twice differentiable cubic B-spline can be derived from (1) and (2):

j	0	1	2	3	4	$\cdots$
$b_{3k\pm j}$	$\frac{4}{6}$	$\frac{4}{6}$	$\frac{2}{6}$	$\frac{1}{6}$	0	0

Hence, $N_k(x)$ is non-zero only for $k - 2 < x < k + 2$.

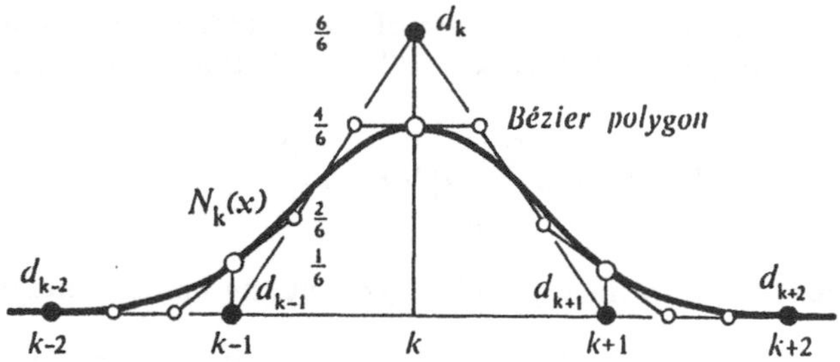

Figure 21.7
Cubic B-spline

Remark 4: The $(m+3)$ B-splines $N_k(x)$, $k = -1, 0, 1, \ldots, m, m+1$ form a basis for the space of all cubic C^2-splines in $[0, m]$.

21.4 The Minimum Property

The curved lines arising from a flexible wooden strip[1] which is held in place by some poles are approximately cubic splines.

Figure 21.8
Spline

Theorem 1: *Cubic C^2-splines minimize the energy integral* $\int_0^m (s''(x))^2 dx$.

For the proof consider the cubic spline $s(x)$ and another twice continuously differentiable function $f(x)$ that coincides with s at all knots and has the same slope as s at the endpoints. Because

$$f(x) = s(x) + (f(x) - s(x)),$$

one gets

$$\int_0^m (f'')^2 dx = \int_0^m (s'')^2 dx + 2\int_0^m s''(f'' - s'')dx + \int_0^m (f'' - s'')^2 dx.$$

Using integration by parts for the second integral on the right side leads to

$$\int_0^m s''(f'' - s'')dx = s''(f' - s')\Big|_0^m - \int_0^m s'''(f' - s')dx.$$

[1] These tools are called splines. They are used by draftsmen in laying out broad sweeping curves.

The first term on the right side vanishes because of the boundary conditions. Since s''' is constant in every interval $[k, k+1]$, say $s''' =: s'''_k$, the second integral has the value

$$\int_0^m s'''(f' - s')dx = \sum_{k=0}^{m-1} s'''_k (f - s)\Big|_k^{k+1} .$$

All the terms of the sum are zero since f and s agree at the knots. Therefore, the sum itself is zero. Further, $\int_0^m (f'' - s'')^2 dx > 0$ for all continuous $f''(x) \neq s''(x)$.

Thus

$$\int_0^m (f''(x))^2 dx \geq \int_0^m (s''(x))^2 dx .$$

Equality is assumed only if $f''(x) = s''(x)$. In this case, $f(x)$, too, equals $s(x)$, as can be seen by integration and the boundary conditions. This then, establishes the theorem.

21.5 B-Splines and Truncated Power Functions

In **21.3** B-splines were derived from piecewise cubic C^2-curves. Moreover, it is also possible to consider B-splines of any general degree n over arbitrarily spaced knots. The basic construction is as follows.

Let pol(x) be a polynomial of degree n and ϕ_i be the truncated power function defined by

$$\phi_i(x) := (x - u_i)_+^n := \begin{cases} 0 & \text{if } x < u_i \\ (x - u_i)^n & \text{if } x \geq u_i \end{cases}$$

where u_i is some given knot. Then pol$(x)+\phi_i(x)$ is in $C^{n-1}(u_i)$, i.e., it is $n-1$-times continuously differentiable at u_i.

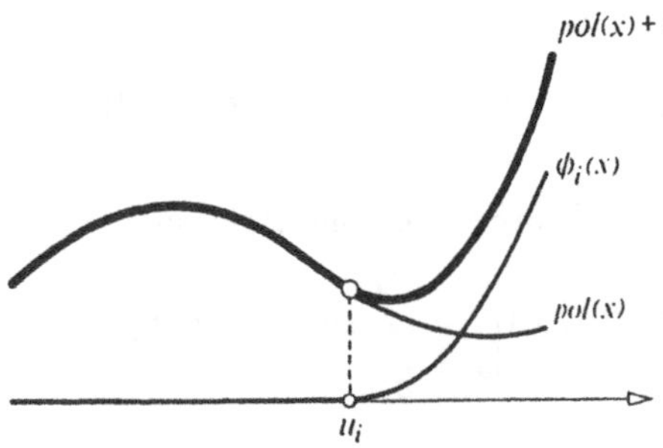

Figure 21.9
C^{n-1} composite polynomials

Let u_i, $i = 0, 1, 2, \ldots$, be some increasing knot sequence, i.e., $u_i < u_{i+1}$. The nth degree B-spline $M_0^n(x)$ over these knots is then defined as a minimal linear combination

$$M_0^n(x) = a_0\phi_0(x) + \cdots + a_m\phi_m(x)$$

where all $a_i \neq 0$ and $M_0^n(x) = 0$ for $x \geq u_m$. Obviously, one has $M_0^n(x) = 0$ for $x < u_0$. The stipulation $M_0^n(x) = 0$ for $x \geq u_{m+1}$ reads after expanding the linear combination above by powers of x

$$\begin{bmatrix} 1 & \cdots & 1 \\ u_0 & & u_m \\ \vdots & & \vdots \\ u_0^n & \cdots & u_m^n \end{bmatrix} \begin{bmatrix} a_0 \\ a_1 \\ \vdots \\ a_m \end{bmatrix} = \begin{bmatrix} 0 \\ 0 \\ \vdots \\ 0 \end{bmatrix},$$

or shorter $U\boldsymbol{a} = \mathbf{o}$.

The homogeneous system $U\boldsymbol{a} = \mathbf{o}$ has non-trivial solutions if and only if $m > n$, cf. **18.1**. The minimum requirement therefore implies $m = n + 1$.

Let U_k denote the $m \times m$-matrix which is obtained from U by deleting its kth column and $\boldsymbol{a} = [\det U_0, \ldots, (-1)^m \det U_m]$. The $(m+1) \times (m+1)$-matrices $V_k = [\boldsymbol{u}_k, U^{\mathrm{t}}]$, $\boldsymbol{u}_k^{\mathrm{t}} = [u_0^k, \ldots, u_m^k]$, are singular, i.e.,

$$\det V_k = \boldsymbol{u}_k^{\mathrm{t}} \boldsymbol{a} = 0\,, \quad k = 0, \ldots, m\,.$$

This means that $\varrho_0 \boldsymbol{a}$ defines the solution of the system $U\boldsymbol{x} = \mathbf{o}$ with the free parameter ϱ_0. Hence $M_0^n(x)$ can be written as

$$M_0^n(x) = \varrho_0 \det \begin{bmatrix} 1 & \cdots & 1 \\ u_0 & \cdots & u_{n+1} \\ \vdots & & \vdots \\ u_0^n & \cdots & u_{n+1}^n \\ \phi_0(x) & \cdots & \phi_{n+1}(x) \end{bmatrix}.$$

Remark 5: The minimum condition entails that the support of $M_0^n(x)$ is $[u_0, u_{n+1}]$.

Remark 6: Analogously, one can construct the functions $M_k^n(x)$. The index k means that $[u_k, u_{k+n+1}]$ is the support of $M_k^n(x)$.

21.6 Normalized B-Splines

Remark 2 in **18.4** suggests to use

$$\varrho_0 = 1/\det \begin{bmatrix} 1 & \cdots & 1 \\ u_0 & \cdots & u_{n+1} \\ \vdots & & \vdots \\ u_0^{n+1} & \cdots & u_{n+1}^{n+1} \end{bmatrix}.$$

Then $M_0^n(x)$ can be considered as the $(n+1)$th divided difference $f_{0,\ldots,n+1}$ of the truncated power function $f(u) = (x-u)_+^n$ with respect to the knots $u_0, \ldots, u_{n+1}$.

Similarly, $M_k^n(x)$ can be defined as the nth divided difference $f_{k,\ldots,k+n+1}$ of $f(u) = (x-u)_+^n$ with respect to $u_k, \ldots, u_{k+n+1}$. Note, that $f_{k,\ldots,k+n+1}$ depends also on x.

One can show that the amplitude of $M_k^n(x)$ decreases considerably with increasing n. This tendency is offset by a normalization. Thus one defines normalized B-splines

$$N_k^n(x) = (u_{k+n+1} - u_k)M_k^n(x).$$

Example 2: The B-splines $N_{k+2}(x)$ introduced in **21.3** are the normalized B-splines $N_k^3(x)$ where $u_k = k$. Note that in **21.3** the index is chosen symmetrically with respect to the support of $N_k^3(x)$.

Example 3: The simplest B-splines are the piecewise constant B-splines $N_k^0(x)$. They are 1 over the interval $[u_k, u_{k+1})$ and 0 elsewhere.

Remark 7: The B-splines $N_k^n(x)$, $k = 0, \ldots, m$, form a basis for all functions in $C^{n-1}[u_0, u_{m+1}]$ which are polynomial over each knot interval $[u_i, u_{i+1}]$. Linear combinations of B-splines are called **splines**.

21.7 De Boor's Algorithm

The above definition of B-splines does not translate into a practical way of computing B-splines. However, it is of fundamental value in deriving B-spline properties from properties about divided differences. For example, one can apply the recurrence relation of divided differences, **18.4**, and the Leibniz formula, **18.12**, to the product $(x-u)(x-u)_+^{n-1}$ and gets the following recurrence relation for B-splines,

$$N_i^n(x) = \alpha_i N_i^{n-1}(x) + (1-\alpha_{i+1})N_{i+1}^{n-1}(x), \tag{3}$$

where $\alpha_i := (x-u_i)/(u_{i+n}-u_i)$ varies from 0 to 1 as x varies form u_i to u_{i+1}.

The recurrence relation (3) is extremely valuable in evaluating a spline

$$s(x) = \sum_{i=0}^{m} d_i^0 \, N_i^n(x)$$

given as a linear combination of B-splines. More than that, relation (3) offers a direct computation of $s(x)$ form the coefficients d_i^0.

If $x \in [u_k, u_{k+1})$, one observes that x lies in the support of only $n+1$ B-splines. Hence one has

$$s(x) = \sum_{i=k-n}^{k} d_i^0 \, N_i^n(x) .$$

The repeated use of the recurrence relation (3) gives

$$s(x) = \sum_{i=k-n+l}^{k} d_i^l \, N_i^{n-l}(x) ,$$

where

(4) $$d_i^l = (1-\alpha_i^l)\, d_{i-1}^{l-1} + \alpha_i^l \, d_i^{l-1} , \qquad \alpha_i^l = \alpha_i^l(x) = \frac{x - u_i}{u_{i+n-l+1} - u_i} ,$$

for $l = 1, 2, \ldots, n$. Because $N_k^0(x) = 1$, one concludes $s(x) = d_k^n$. This method of computing $s(x)$ is known as **de Boor's algorithm.**

If the intermediate coefficients are stored in a Neville-like scheme

$$\begin{array}{llll} d_{k-n}^0 & & & \\ d_{k-n+1}^0 & d_{k-n+1}^1 & & \\ \vdots & \vdots & \ddots & \\ d_k^0 & d_k^1 & \cdots & d_k^n = s(x) \end{array}$$

one may present computation rule (4) graphically as

$$\begin{array}{ll} d_{i-1}^{l-1} & \searrow \\ d_i^{l-1} & \longrightarrow d_i^l . \end{array}$$

Note that the α_i^l depend on x and n.

Remark 8: In general, the intermediate B-splines $N_i^{n-l}(x)$ are not evaluated. However, if they are stored in a scheme similar to that of the coefficients α_i^l

$$\begin{array}{llll} N_{k-n}^n & & & \\ N_{k-n+1}^n & N_{k-n+1}^{n-1} & & \\ \vdots & \vdots & \ddots & \\ N_k^n & N_k^{n-1} & \cdots & N_k^0 = 1 \end{array}$$

the influence of N_i^{n-l} on N_{i-1}^{n-l+1} and N_i^{n-l+1} can be presented graphically as

$$\begin{array}{ll} N_{i-1}^{n-l+1} & \nwarrow \\ N_i^{n-l+1} & \leftarrow N_i^{n-l}. \end{array}$$

Remark 9: Often one deals with parametric spline curves

$$s(x) = \sum_{i=0}^{m} d_i \, N_i^n(x)$$

with vector-valued coefficients $d_i \in \mathbb{R}^m$, $m \geq 2$. Such a curve can be evaluated in exactly the same way as a spline function $s(x)$. More than that one gets a geometric construction defined by (4). It is similar to de Casteljau's construction in **20.3**. Figure 21.10 shows this construction for a cubic spline in $\mathbb{R}^3$.

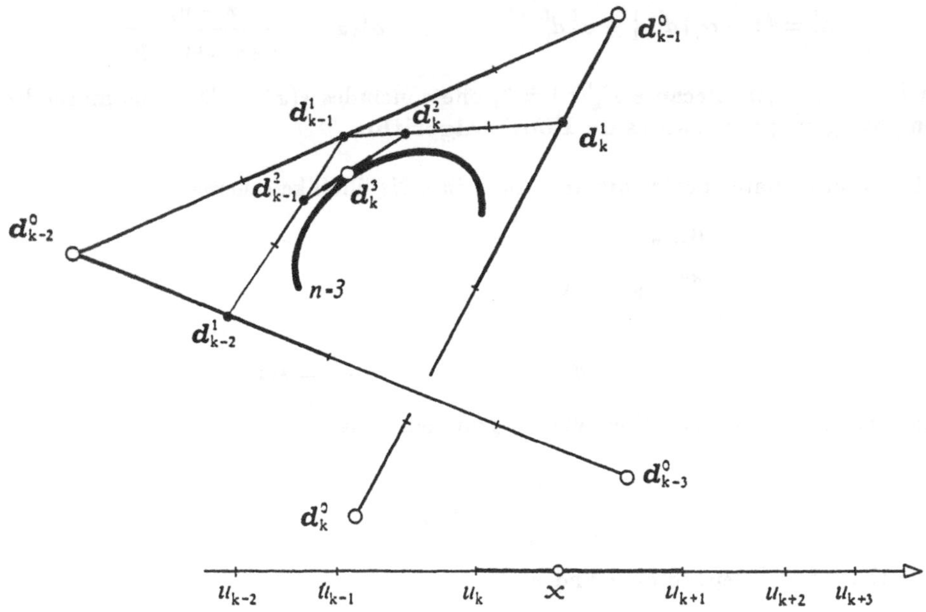

Figure 21.10 De Boor's algorithm

21.8 Notes and Exercises

1. Every sequence of increasingly finer segmentations of an arbitrary function $f(x)$ defines a sequence of spline interpolants of fixed degree n which converges to

$f(x)$ if $f(x)$ is continuous and if the distance between successive knots goes to zero.

2. Use **20.5** number 6 to show that the radii of curvature of two spline segments at a common knot are identical.

3. B-splines can be defined also for coalescing knots as in **21.5** if one uses additional truncated power functions of lower degree.

4. The normalized B-splines can be viewed as generalizations of (truncated) Bernstein polynomials. For example one has

$$\sum_{i=0}^{m} N_i^n(x) = 1 \text{ for all } x \in [u_n, u_{m+1}) ,$$

$$N_i^n(u) > 0 \text{ for all } x \in (u_i, u_{i+n+1}) .$$

5. A **bispline** is a segmented surface such that both families of parameter lines are splines.

V Numerical Differentiation and Integration

In many applications one resorts to numerical differentiation and numerical integration particularly to solve differential equations. An intrinsic difficulty, then, is to solve a continuous problem by the discrete arithmetic of a computer and to represent its solution by finitely many numbers. The initial value problem for ordinary differential equations provides a good example of how to tackle this problem.

22 Numerical Differentiation and Integration

Numerically, one determines the derivatives of a function $f(x)$ from a function $a(x)$ which approximates $f(x)$ and whose derivatives are easy to compute. In a similar manner, one can determine the integral of $f(x)$. Usually one splits the function $f(x)$ into segments for this purpose.

22.1 Numerical Differentiation

A simple strategy is to interpolate $f(x)$ by a polynomial of degree n in a neighborhood of the point x at which one seeks the derivative of $f(x)$ and to differentiate this polynomial instead. Assuming equidistant interpolation abscissae $x_i := x_0 + ih$, $i = 0, 1, \ldots, n$, one gets with the notation of **18.1**

for $n = 1$

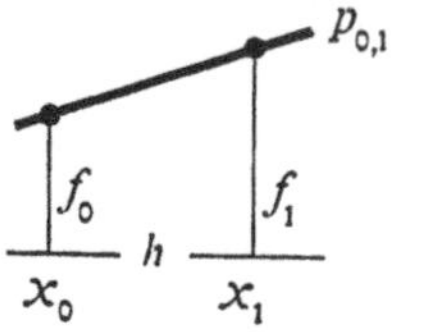

$$f' = \frac{1}{h}(f_1 - f_0)$$

and for $n = 2$

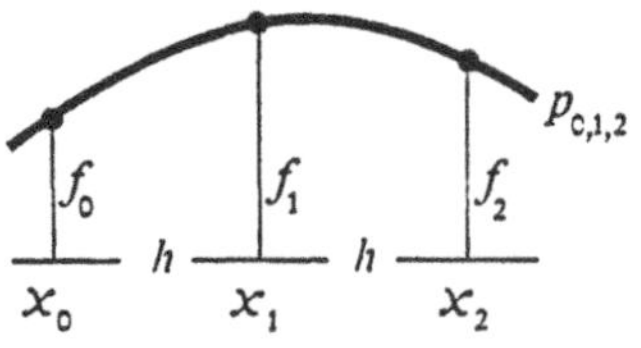

$$f_0' = \frac{1}{2h}(-3f_0 + 4f_1 - f_2)$$

$$f_1' = \frac{1}{2h}(f_2 - f_0)$$

$$f_2' = \frac{1}{2h}(f_0 - 4f_1 + 3f_2)$$

$$f'' = \frac{1}{h^2}(f_0 - 2f_1 + f_2)$$

etc. The formulas are all of the form

$$\sum_{i=0}^{n} \beta_i f_i\,, \text{ with } \sum_{i=0}^{n} \beta_i = 0\,.$$

In labeling these formulas it is expedient to display them as shown below. The weights for the interpolation values appear underneath the interpolation abscissae, the denominators at their right side. Each row of the table describes derivatives of the same order which is indicated in the leftmost column. The point at which the derivative is taken is marked by ○.

Differentiation

f'	-1	1	h	-1	1	h	-1	1	h

f'	-3	4	-1	$2h$	-1	0	1	$2h$	1	-4	3	$2h$
f''	1	-2	1	h^2	1	-2	1	h^2	1	-2	1	h^2

f'	-11	18	-9	2	$6h$	1	-27	27	-1	$24h$	-2	9	-18	11	$6h$
f''	2	-5	4	-1	h^2	1	-1	-1	1	$2h^2$	-1	4	-5	2	h^2
f'''	-1	3	-3	1	h^3	-1	3	-3	1	h^3	-1	3	-3	1	h^3

22.2 Error Estimates for the Numerical Differentiation

The deviation $P_{n+1}(x) = f(x) - \mathrm{pol}(x)$ of the interpolation polynomial from the function $f(x)$ was shown in **19.2** to be

$$P_{n+1}(x) = \frac{1}{(n+1)!} f^{(n+1)}(y) \prod_{i=0}^{n} (x - x_i) \, .$$

On differentiating this equation with respect to x one obtains for $x = x_k$

$$f'(x_k) - \mathrm{pol}'(x_k) = \frac{1}{(n+1)!} f^{(n+1)}(y(x_k)) \prod_{\substack{i=0 \\ i \neq k}}^{n} (x_k - x_i) \, ,$$

and thence for equidistant support abscissae $x_i = x_0 + ih$

$$|f'(x_k) - \mathrm{pol}'(x_k)| \leq M h^n \, ,$$

where M is some constant with

$$M \geq \left|\frac{1}{n+1} f^{(n+1)}(x)\right| \text{ for all } x \in [x_0, x_n].$$

This means that the error is asymptotically bounded by some multiple of h^n as h goes to zero. $\mathrm{pol}'(x_k)$ is therefore called an **approximation of order n** for $f'(x_k)$.

Moreover, a general function $A(h)$ is said to be an approximation of **order p** for the value a if p is the largest number such that there is a constant M with

$$|A(h) - a| \leq M h^p \text{ for all small } h.$$

Example 1: $\frac{1}{h}(f(x+h) - f(x))$ is an approximation of order one for $f'(x)$. This follows immediately from the Taylor expansion of f about x.

Example 2: $\frac{1}{2h}(f(x+h) - f(x-h))$ is an approximation of order two for $f'(x)$. This follows directly from the Taylor expansion of f about x.

Remark 1: Often, the symbol O is used to define the order of convergence: The notation

$$A(h) = a + O(h^p)$$

is used to express that $A(h)$ is an approximation of order p for a. This means that $O(h^p)$ stands for a function for which $\dfrac{O(h^p)}{h^p}$ is bounded as $h \to 0$.

Remark 2: Raising the degree of the interpolation polynomial alone does, in general, if h is fixed, not improve the result.

22.3 Numerical Integration

An approximation F for the integral of $f(x)$ is obtained similarly through integration of the interpolation polynomial $p_{0,\ldots,n}$. On using the notation of **18.1** and the notation l for the length of the integration interval one gets in case of equidistant support abscissae $x_i = x_0 + ih$:

the **Trapezoidal rule** for $n = 1$:

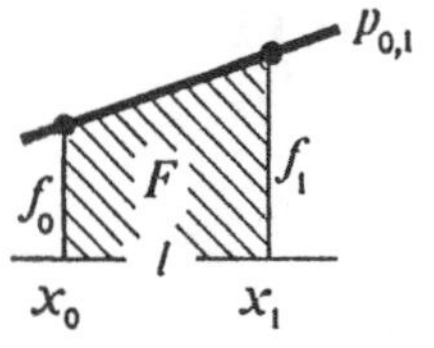

$$F = \frac{l}{2}(f_0 + f_1),$$
$$l = h,$$

Simpson's rule for $n = 2$:

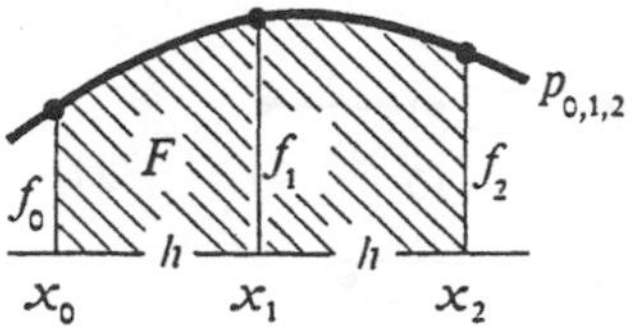

$$F = \frac{l}{6}(f_0 + 4f_1 + f_2),$$
$$l = 2h,$$

Bessel's rule for the inner segment and $n = 3$:

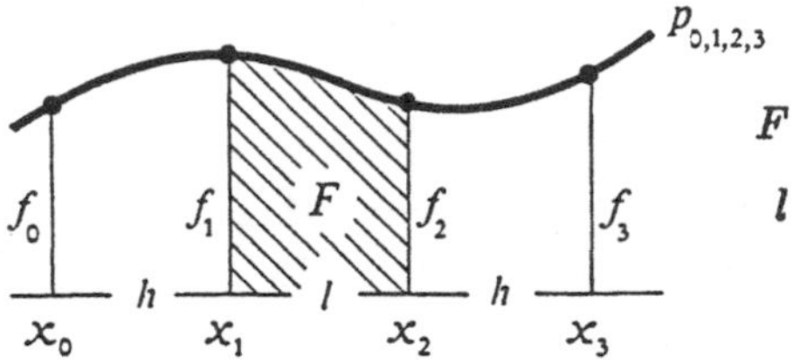

$$F = \frac{l}{24}(-f_0 + 13f_1 + 13f_2 - f_3),$$
$$l = h,$$

or the so-called **Midpoint rule** for $n = 0$:

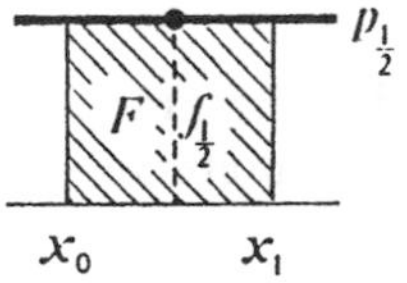

$$F = l\,f_{1/2},$$
$$l = h$$

where $f_{1/2}$ denotes the value of f at $x_0 + \frac{1}{2}h$.

All these formulas can be written as

$$\int_a^b p_{0,\ldots,n}(x)dx = l \cdot \sum_{i=0}^{n} \alpha_i f_i$$

where $\sum_{i=0}^{n} \alpha_i = 1$; they are called **Newton-Cotes** formulas.

A similar notation as in **22.1** can be used to present these formulas. The weights appear underneath the equidistant support abscissae. They have to be divided by their sum which is written at the right side. The integration interval is marked by  . For $l = 1$, these formulas are:

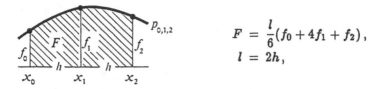

Midpoint

1 1

Trapezoid

1 1 2

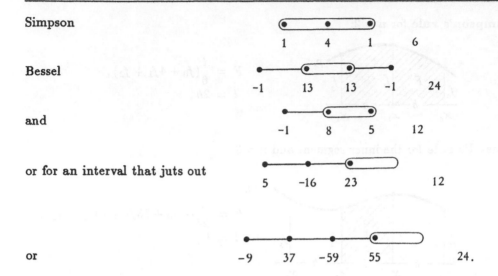

Simpson

Bessel

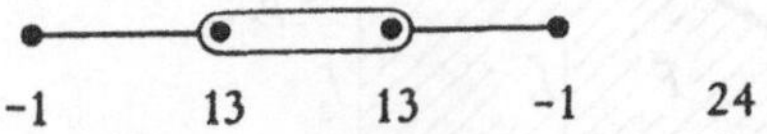

and

or for an interval that juts out

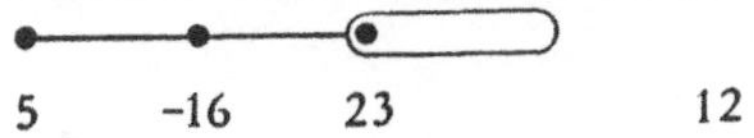

or

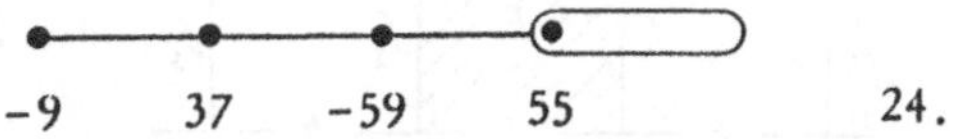

22.4 Composite Integration Rules

Large integration intervals $I = [a, b]$ should be subdivided for practical reasons. An obvious approach is to split I at $x_k = a + kh$ into m subintervals of uniform width $h = \frac{b-a}{m}$.

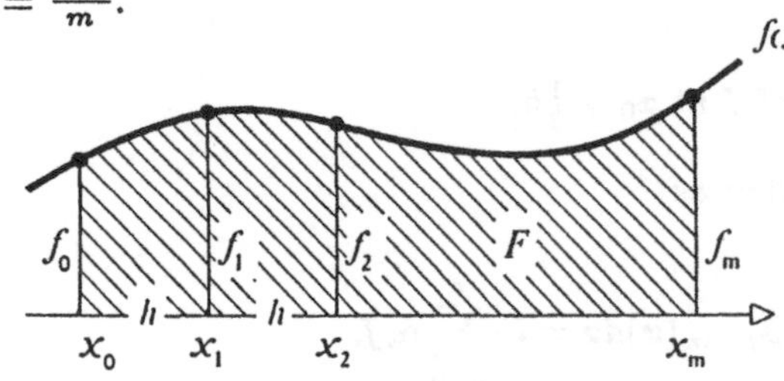

Figure 22.1
Segmentation

Then a simple integration rule can be used for each segment. Their "sum" is a so-called **composite integration rule** for the approximation of the integral

$$\int_a^b f(x)dx\,,$$

e.g., the **Composite Midpoint rule**

$$M := h(f_{1/2} + f_{3/2} + f_{5/2} + \cdots + f_{m-1/2})\,,$$

the **Composite Trapezoid rule**

$$T := \frac{h}{2}(f_0 + 2f_1 + \cdots + 2f_{m-1} + f_m)\,,$$

or, in case of even m, the **Composite Simpson's rule**

$$S := \frac{h}{3}(f_0 + 4f_1 + 2f_2 + 4f_3 + \cdots + 4f_{m-1} + f_m).$$

22.5 Error Estimation for the Numerical Integration

The deviation $P_{n+1}(x) = f(x) - \mathrm{pol}(x)$ of a single segment k with the interpolation abscissae $z_0, \ldots, z_n$ is integrated similar to **22.2** where its derivative is considered

$$(1) \quad \int_{x_k}^{x_{k+1}} f(x)dx - \int_{x_k}^{x_{k+1}} \mathrm{pol}(x)dx = \frac{1}{(n+1)!} \int_{x_k}^{x_{k+1}} f^{(n+1)}(y(x)) \prod_{j=0}^{n}(x - z_j)dx .$$

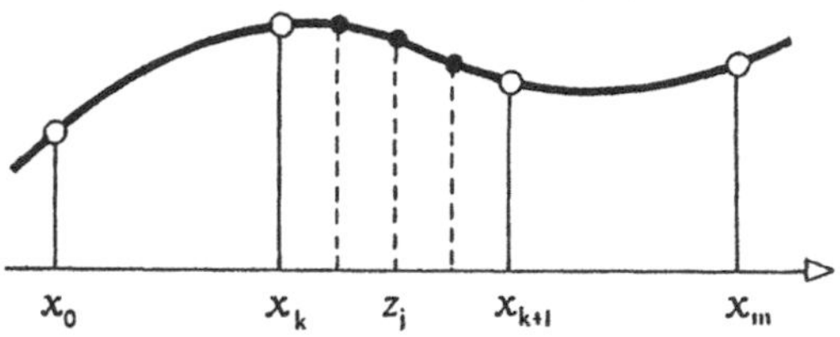

Figure 22.2
Interpolation abscissae of a segment

For $n = 1$, the trapezoid rule, one has $(x - x_k)(x - x_{k+1}) \leq 0$ in $[x_k, x_{k+1}]$. Hence, the mean-value theorem for integrals can be used to simplify the error formula, namely

$$\frac{1}{2} \int_{x_k}^{x_{k+1}} f''(y(x))(x - x_k)(x - x_{k+1})dx = -\frac{h^3}{12} f''(y_k) \text{ where } y_k \in [x_k, x_{k+1}] .$$

Finally, a summation over all segments gives

$$\int_a^b f(x)dx - T = -\frac{h^3}{12} \sum_{k=0}^{m-1} f''(y_k) = -\frac{h^2}{12}(b - a)f''(y) .$$

Thus, the composite trapezoid rule is of order 2.

Remark 3: The composite midpoint rule is of the same order 2, whereas the composite Simpson's rule is of order 4.

If $z_0, \ldots, z_n$ are equidistant, if n is even, and if f is a polynomial of degree $\leq n+1$, then $f^{(n+1)}(x)$ is constant and

$$\int_{x_k}^{x_{k+1}} \prod_{j=0}^{n}(x - z_j)dx = 0$$

because of symmetry reasons.

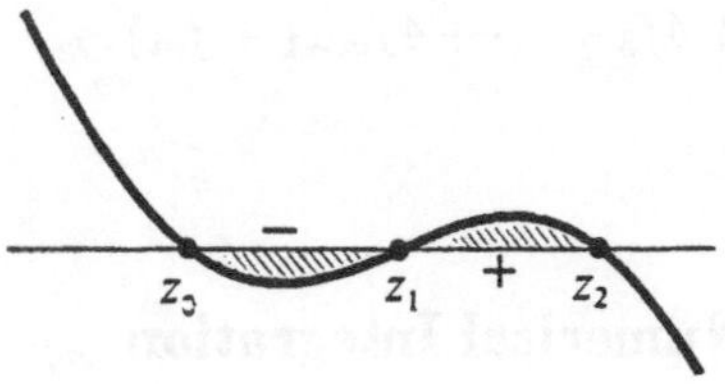

Figure 22.3
Odd degree polynomial $\prod_{j=0}^{n}(x - z_j)$

This means in view of (1) that polynomials of degree $n+1$, n even, are exactly integrated by only an n-point integration formula.

Example 3: The four-point integration rule given graphically as in **22.3**

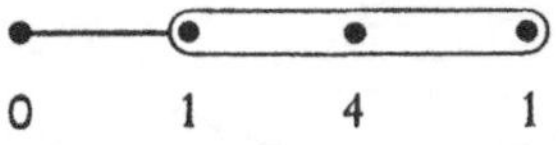

must equal Simpson's rule.

22.6 Notes and Exercises

1. The formulas for the numerical integration and differentiation can easily be checked for $f = 1, x, x^2$, etc.

2. The integral of the Bézier polynomial $\text{pol}(\lambda) = \sum_{i=0}^{n} b_i B_i^n(\lambda)$ can be derived from **20.2**. One gets

$$\int_0^1 \text{pol}(\lambda) d\lambda = \frac{1}{n+1} \sum_{i=0}^{n} b_i$$

and in particular $\int_0^1 B_i^n(\lambda) d\lambda = \frac{1}{n+1}$.

3. For the cubic splines in **21.3** one gets $(m \geq 2)$

$$\int_0^m s(x)dx = \frac{1}{4}(b_0 + b_1 + 4b_3 + \cdots + 4b_{3m-3} + b_{3m-1} + b_{3m}).$$

4. For the twice differentiable cubic splines of **21.3** one gets $(m \geq 3)$

$$\int_0^m s(x)dx = \frac{1}{12}(3b_0 + 4d_0 + 11d_1 + 12d_2 + \cdots + 12d_{m-2}$$

$$+ 11d_{m-1} + 4d_m + 3b_{3m}).$$

23 Extrapolation

If some value has been determined by an approximation one can try to obtain approximations of higher order by **extrapolation.** Aitken's Δ^2-process in **13.4** is a discrete example of this idea.

23.1 Sequences of Approximations

If an approximation $A(h)$ for some number a is continuous in h at $h = 0$ and assumes the value a at $h = 0$, then one gets for every zero sequence $h_0, h_1, \ldots$ a sequence of approximations $A_0, A_1, \ldots$ which goes to a. Often a **geometric sequence of step sizes** is chosen

$$h_k := q^k h_0 \text{ with } 0 < q < 1 .$$

Example 1: Consider the composite trapezoid rule for approximating the integral of $f(x)$ over the interval $[a, b]$ together with the step sizes $h_k := (b - a)/2^k$, i.e., $q = 1/2$. The following algorithm calculates the respective approximations T_k, $k = 0, \ldots, N$, taking into account that T_{k-1} is part of the sum T_k.

Composite Trapezoid Rule

Input: $f(x)$; a, b; N

Output: $T_0, \ldots, T_N$ (trapezoid sums)

1 Set $h := b - a$

2 and $T_0 := \frac{h}{2}(f(a) + f(b))$.

3 For $k = 1, 2, \ldots, N$

4 set $h := \frac{h}{2}$,

5 determine $M_{k-1} := 2h \sum_{i=1}^{2^{k-1}} f(a + (2i - 1)h)$

6 and $T_k := \frac{1}{2}(T_{k-1} + M_{k-1})$.

Remark 1: The M_{k-1} in line **5** are the midpoint rule approximations of **22.4**.

23.2 Richardson Extrapolation

Often the convergence of the sequence $A_k := A(h_k)$ can be accelerated. Namely, if the approximation $A(h)$ has the expansion

$$A(h) = a + a_p h^p + O(h^{p+r})$$

with $a_p \neq 0$, then $A(h)$ is of order p and one can proceed as in **13.4**.

The term $a_p h^p$ can be eliminated from two approximations

$$A_0 = A(h_0) = a + a_p h_0^p + O(h_0^{p+r}),$$
$$A_1 = A(h_1) = a + a_p h_1^p + O(h_1^{p+r}).$$

With $h_1 = q h_0$ one gets

$$a = \frac{A_1 - q^p A_0}{1 - q^p} + O(h_0^{p+r}).$$

Hence,

$$A_{0,1} := \frac{A_1 - q^p A_0}{1 - q^p}$$

is an approximation for a of order $p + r$ at least.

Remark 2: The process of finding $A_{0,1}$ from A_0 and A_1 can be thought of as **extrapolation** from the two points (h_0, A_0), (h_1, A_1) at $h = 0$ by a polynomial of the form $c_0 + c_p h^p$. The coefficients $c_0 = A_{0,1}$ and $c_p = \dfrac{A_1 - A_0}{h_0^p(q^p - 1)}$ approximate a and a_p.

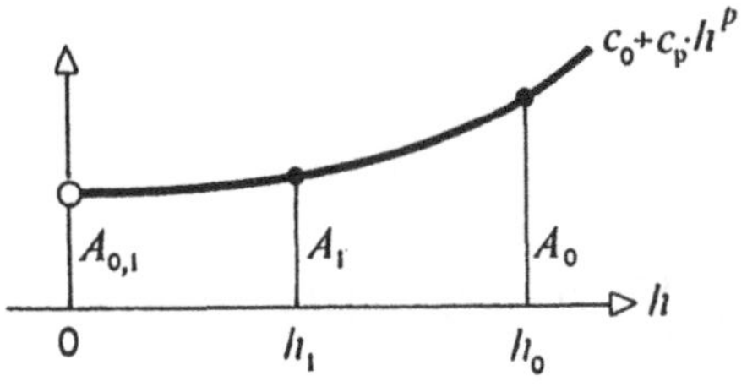

Figure 23.1
Extrapolation

For any geometric sequence of step sizes $h_k := q^k h_0$ with the corresponding approximations $A_k := A(h_k)$ of order p, one can define the sequence

$$A_{k-1,k} := \frac{A_k - q^p A_{k-1}}{1 - q^p}$$

in analogy to the process of obtaining $A_{0,1}$ from A_0 and A_1. The $A_{k-1,k}$ are approximations of order $p + r$ at least.

Remark 3: In the following sense, the sequence $A_{k-1,k}$ converges **faster** towards a than the sequence A_k:

$$\frac{A_{k-1,k}-a}{A_k-a}=O(h^r).$$

Example 2: Consider the **symmetric difference quotient**

$$D(h):=\frac{1}{2h}(f(x+h)-f(x-h)).$$

From the Taylor expansion of $f(x\pm h)$ around x one gets

$$D(h)=f'(x)+\frac{1}{3!}f'''(x)h^2+\frac{1}{5!}f^{(5)}(x)h^4+\cdots.$$

Hence, $D(h)$ is of order 2 if $f'''(x)\neq 0$, and Richardson's extrapolation would provide approximations of order $O(h^4)$.

Example 3: The trapezoidal sum has an expansion of the form

$$T(h)=\int_a^b f(x)dx+s_2h^2+s_4h^4+\cdots$$

with $h=\frac{b-a}{m}$, and any integer m. Therefore it is of order 2. The proof is omitted here. Extrapolation with $q=\frac{1}{2}$ yields

$$T_{k-1,k}:=\frac{T_k-\frac{1}{4}T_{k-1}}{1-\frac{1}{4}}=\frac{1}{3}(4T_k-T_{k-1}).$$

Remark 4: One can easily convince oneself that the $T_{k-1,k}$ in Example 3 above are the composite Simpson's rule approximations for the step sizes $\frac{h_0}{2},\frac{h_0}{4},\frac{h_0}{8},\ldots$.

23.3 Iterated Richardson Extrapolation

Obviously, one can try to repeat the extrapolation. This is feasible and especially simple if the approximation $A(h)$ has an asymptotic expansion of the form

$$A(h)=a+a_ph^p+a_{2p}h^{2p}+\cdots+a_{np}h^{np}+O(h^{np+r})$$

with non-vanishing coefficients a_{ip}. Inserting this into the extrapolation rule gives the expansion

$$A_{k-1,k}=\frac{A_k-q^pA_{k-1}}{1-q^p}=a+b_{2p}h^{2p}+\cdots+O(h^{np+r})$$

with new coefficients b_{ip} where $h := h_{k-1}$. This shows that $A_{k-1,k}$ is of order $2p$. The **iterated Richardson extrapolation** generates the approximations

$$A_{k-2,k-1,k} := \frac{A_{k-1,k} - q^{2p} A_{k-2,k-1}}{1 - q^{2p}}$$

of order $3p$ and so forth up to order $np + r$. The general recursion is of the form

$$A_{i,\ldots,k} := \frac{A_{i+1,\ldots,k} - q^{(k-i)p} A_{i,\ldots,k-1}}{1 - q^{(k-i)p}} .$$

It is convenient to organize the respective calculations in a Neville-like **scheme for the iterated Richardson extrapolation**

	1	q^p	q^{2p}		$\cdots$	q^{np}
	A_0		$k-i$			
i	A_1	$A_{0,1}$				
	A_2	$A_{1,2}$	$A_{0,1,2}$			
k	A_3	$A_{2,3}$	$A_{1,2,3}$	$A_{0,1,2,3}$		
	$\vdots$	$\vdots$	$\vdots$	$\vdots$	$\ddots$	
	A_n	$A_{n-1,n}$	$A_{n-2,n-1,n}$		$\cdots$	$A_{0,\ldots,n}$
	$\vdots$	$\vdots$	$\vdots$	$\vdots$		$\vdots$

The $q^{(k-i)p}$ are written above the columns $k - i$ for convenience.

Remark 5: Each column converges faster in the sense of Remark 3 than the preceding one.

Example 4: The Romberg integration in the following section is an example of the iterated Richardson extrapolation.

23.4 Romberg Integration

The asymptotic expansion of the trapezoidal sum in Example 3 allows for an unlimited iteration of Richardson's extrapolation.

The first column of the corresponding Neville like scheme contains the trapezoid sums T_k, the second column the Simpson sums. The algorithm below, however, computes the scheme row by row with $q = \frac{1}{2}$ where the $T_{k-1,k}, \ldots, T_{0,\ldots,k}$ are written over the $T_{k-1}, \ldots, T_0$. It is easier, then, to add another row for a further halving of the step size.

This procedure was first proposed by Romberg:

Romberg Integration

Input: $T_0, \ldots, T_N$ (trapezoid sums)
Output: $T := T_{0,\ldots,N}$

For $k = 1, 2, \ldots, N$
  set $r := 1$.
  For $i = k-1, k-2, \ldots, 0$
    set $r := 4r$
    determine $T_i := \dfrac{rT_{i+1} - T_i}{r-1}$.
Set $T := T_0$.

23.5 Notes and Exercises

1. An asymptotic expansion

$$A(h) = a + a_1 h + a_2 h^2 + \cdots + a_p h^p + O(h^{p+1})$$

which exists for every p does not necessarily converge as a series.

2. One can obtain the recursion formula for the $A_{i,\ldots,k}$ of the Richardson extrapolation from Aitken's Lemma by setting $p_{i,\ldots,k} := A_{i,\ldots,k}$ and $x := 0$, $x_i := q^i h_0$, $x_k := q^k h_0$.

3. Because of 2, one can interpret the determination of the $A_{i,\ldots,k}$ as **iterated linear extrapolation** where the A_k are associated with the abscissae q^{kp} as opposed to the abscissae h^k in Remark 3.

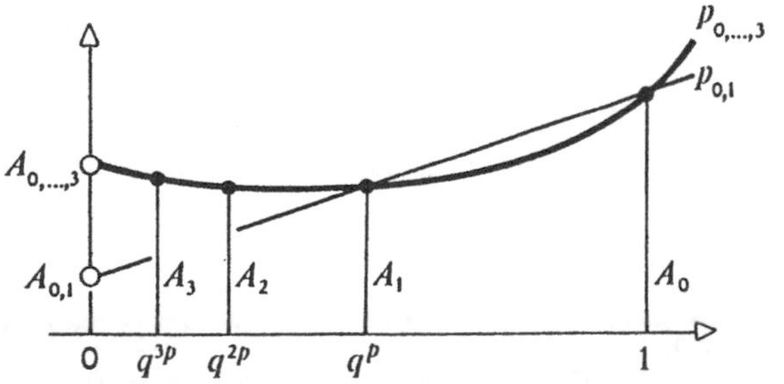

Figure 23.2
Iterated linear extrapolation

4. The extrapolation generates the solution only for sufficiently small step sizes h. There are approximations with asymptotic expansions for which the extrapolation is impractical.

5. Already the fourth column of the Romberg scheme cannot be obtained through polynomial interpolation as are the Simpson sums of the second column.

24 One-Step Methods for Differential Equations

The problem of finding the solution to a **differential equation** or system of differential equations of **order 1**

$$y' = f(x, y)$$

in an interval $[a, b]$ with the constraint $y_0 = y(a)$ is called **initial-value problem.** Numerically it is solved by using repeatedly a discretized version of the differential equation.

24.1 Discretization

A first-order differential equation[1)]

$$y' = f(x, y)$$

with a continuous function f over $I := [a, b] \times \mathbb{R}$ defines a **vector field.**

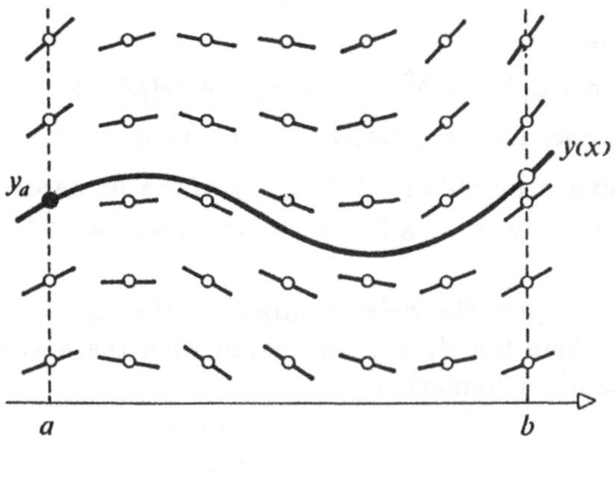

Figure 24.1
Vector field

By a theorem from the theory of differential equations there is a unique solution $y(x)$ for every **initial value** y_0 if f satisfies a **Lipschitz condition**

$$|f(x, y) - f(x, z)| \leq L|y - z|.$$

The basic idea of a numerical solution is to subdivide the interval $[a, b]$ by some (for the moment equidistant) **meshpoints** $x_k := a + kh$, $k = 0, \ldots, m$, and to replace the solution $y(x)$ by a polynomial approximation, e.g., a polygonal line, i.e., piecewise linear approximation $\eta(x)$ with the vertices (x_k, η_k) where $\eta_0 := y_0$.

---

1) For systems of equations, y and y' are coordinate columns; then $|\cdot|$ has to be interpreted as $\|\cdot\|$.

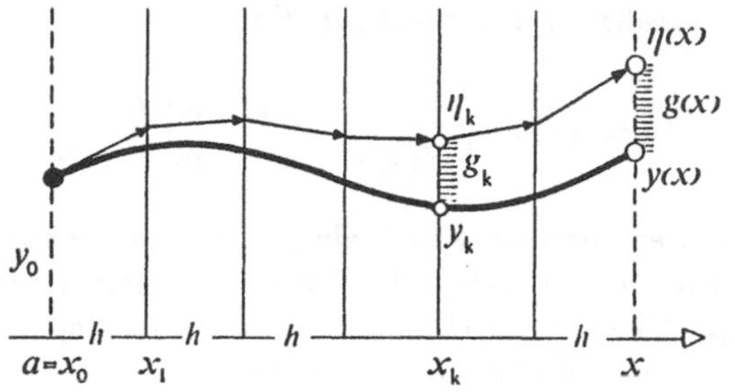

Figure 24.2 Approximation by a polygonal line and global discretization error

The simplest construction of such a polygonal or piecewise linear approximation goes back to Euler. He used

$$\eta_{k+1} := \eta_k + hf(x_k, \eta_k)\,.$$

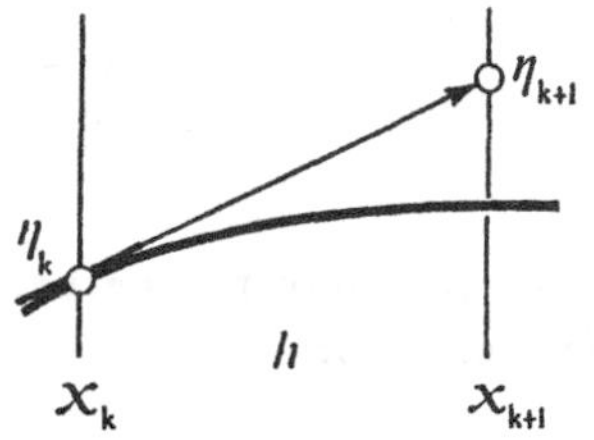

Figure 24.3
Euler's method

More generally, a one-step method is defined by the construction

$$\eta_{k+1} := \eta_k + hF(x_k, \eta_k)$$

where the **increment function** F depends also on f (as well as on x_k, η_k) and usually also on h.

24.2 Discretization Error

It is reasonable to require that the approximation $\eta(x)$ produced by a one-step method with step size $h = \frac{x-a}{m}$ converges to $y(x)$ as m becomes large for each $x \in [a, b]$. A one-step method is said to be **convergent of order p** if

$$\eta(x) - y(x) = O(h^p)\,.$$

The difference

$$g_k := \eta_k - y_k\,, \quad k = 0, 1, \ldots, m\,,$$

is called **global discretization error** at $x = x_k$.

With the ideal (but idealistic) increment function

$$D(x_k, y_k) = \begin{cases} \frac{1}{h}(y_{k+1} - y_k) & \text{for } h > 0 \\ y'(x_k) & \text{for } h = 0 \end{cases}$$

instead of F a one-step method would always generate exact values $\eta_k = y_k$, $k = 0, 1, \ldots, m$. However, on using the function F errors are potentially committed at each step. The error $l_k(h)$ which is made in step k if η_k happened to equal y_k is called the **local discretization error**, i.e.,

$$l_k(h) := D(x_k, y_k) - F(x_k, y_k).$$

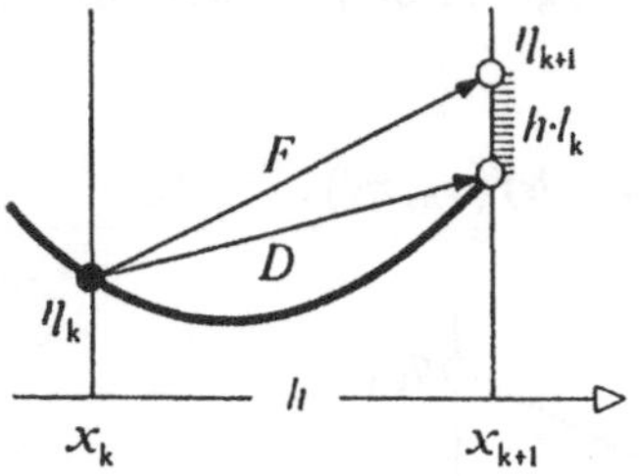

Figure 24.4
Local discretization error

The following theorem shows that the convergence behavior of g_k is essentially inherited from the local discretization error:

Theorem 1: *If the local discretization error of a one-step method is of order $p \geq 1$ and if the increment function F is Lipschitz continuous in its second argument then the one-step method is convergent of order at least p.*

For the proof, one starts out with the equations

$$\eta_{k+1} = \eta_k + hF(x_k, \eta_k) \quad \text{and} \quad y_{k+1} = y_k + hD(x_k, y_k)$$

whence

$$g_{k+1} = g_k + h[D(x_k, y_k) - F(x_k, y_k) + F(x_k, y_k) - F(x_k, \eta_k)].$$

On employing the inequalities

$$|F(x_k, y_k) - F(x_k, \eta_k)| \leq L|y_k - \eta_k|$$

and

$$|D(x_k, y_k) - F(x_k, y_k)| \leq Mh^p$$

with suitable constants L and M, the estimate

$$|g_{k+1}| \leq (1 + hL)|g_k| + Mh^{p+1}$$

follows. Since $g_0 = 0$, a recursive insertion yields

$$|g_{k+1}| \leq [(1+hL)^k + \cdots + (1+hL) + 1]Mh^{p+1}$$
$$= \frac{(1+hL)^{k+1}-1}{L}Mh^p .$$

Since $1 + hL \leq e^{hL}$, one finally gets for $k+1 = m$ and $mh = x - a$

$$|g(x)| \leq \frac{e^{(x-a)L}-1}{L}Mh^p .$$

This verifies Theorem 1.

Example 1: For **Euler's method** with $F = f(x_k, \eta_k)$ one has $p = 1$. This can be seen by the Taylor expansion of the solution $y(x)$ at x_k,

$$D(x_k, y_k) - F(x_k, y_k) = \frac{1}{2}y''h + \dots .$$

Hence, $l_k(h) = O(h)$ and by Theorem 1 also $g(x) = O(h)$.

Example 2: **Heun's method** is given by

$$F(x_k, \eta_k) = \frac{1}{2}(f_0 + f_1)$$

where

$$f_0 := f(x_k, \eta_k),$$
$$f_1 := f(x_k + h,\ \eta_k + hf_0).$$

This method is of order ≥ 2 for every differential equation as is shown in Example 3.[1)]

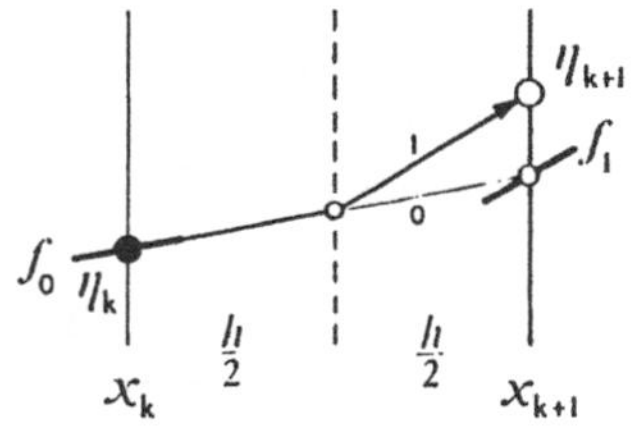

Figure 24.5
Heun's method

24.3 Runge-Kutta Methods

Euler's and Heun's method are special cases of the general **Runge-Kutta methods**. Those are one-step methods of the following form:

[1)] The small numbers in the figure indicate the order of the construction.

To $n+1$ functional values

$$\begin{aligned} f_0 &:= f(x_k, \eta_k)\,, \\ f_1 &:= f(x_k + \alpha_1 h,\ \eta_k + \beta_1 h)\,, \\ &\ \vdots \\ f_n &:= f(x_k + \alpha_n h,\ \eta_k + \beta_n h)\,, \end{aligned}$$

and

$$\alpha_i = \sum_{j=0}^{i-1} \beta_{i,j}\,, \qquad \beta_i = \sum_{j=0}^{i-1} \beta_{i,j} f_j$$

one defines

$$F := \gamma_0 f_0 + \gamma_1 f_1 + \cdots + \gamma_n f_n\,.$$

The so-called **Butcher array** below lends itself to a convenient notation of the coefficients $\beta_{i,j}$, α_i and γ_j.

$$\begin{array}{c|cccc} 0 & 0 & & \cdots & 0 \\ \alpha_1 & \beta_{1,0} & 0 & & \\ \vdots & \vdots & & \ddots & \vdots \\ \alpha_n & \beta_{n,0} & \cdots & \beta_{n,n-1} & 0 \\ \hline & \gamma_0 & \cdots & \gamma_{n-1} & \gamma_n \end{array}$$

The matrix of the $\beta_{i,j}$ is non-zero only below its diagonal. This means that only the f_j, $j = 0, 1, \ldots, i-1$, calculated prior to f_i are needed to compute f_i. These Runge-Kutta methods are therefore said to be **explicit**.

The coefficients $\beta_{i,j}$ and γ_j for some fixed n are chosen so that the local discretization error is of a certain order p. This set-up is deduced from the Taylor expansion of $F(x_k, \eta_k)$ as a function of h about $h = 0$. This leads to a system of non-linear equations for the α_i, $\beta_{i,j}$ and γ_j. Each solution defines a specific one-step method.

Example 3: Let $y(x)$ be the solution to the initial values x_k, η_k. Then the Taylor expansion of $F = \gamma_0 f_0 + \gamma_1 f_1$ at x_k, y_k about $h = 0$ is as follows:

$$\begin{aligned} F &= \gamma_0 f + \gamma_1 (f + h(\alpha_1 \frac{\partial f}{\partial x} + \beta_{1,0} \frac{\partial f}{\partial y} f) + O(h^2)) \\ &= (\gamma_0 + \gamma_1) y' + h \gamma_1 \alpha_1 y'' + O(h^2) \end{aligned}$$

since $\alpha_1 = \beta_{1,0}$. This is to be compared with the Taylor expansion of D at x_k, y_k about $h = 0$,

$$D = y' + \frac{1}{2} y'' h + O(h^2)\,.$$

If $p = 2$ is desired, i.e., $D - F = O(h^2)$, this leads to the conditions

$$\gamma_0 + \gamma_1 = 1, \quad \gamma_1 \alpha_1 = \frac{1}{2}.$$

The solutions form a one-parameter family of Runge-Kutta methods which are all of order ≥ 2.

Among these solutions are Heun's method of Example 2 where

$$\gamma_0 = \gamma_1 = \frac{1}{2} \quad \text{and} \quad \alpha_1 = \beta_{1,0} = 1$$

as well as the **Midpoint method** with

$$\gamma_0 = 0, \quad \gamma_1 = 1 \quad \text{and} \quad \alpha_1 = \beta_{1,0} = \frac{1}{2}.$$

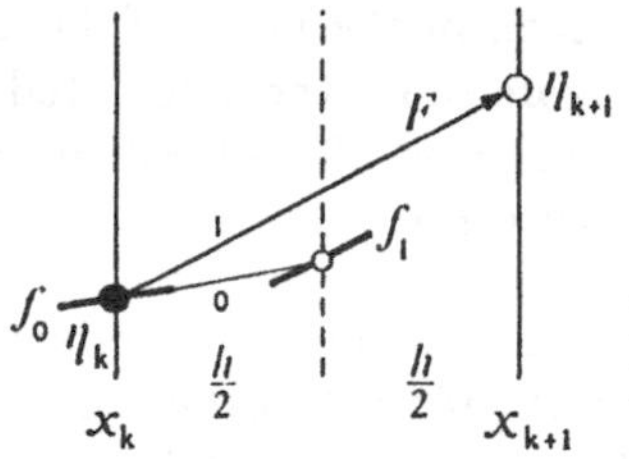

Figure 24.6
Midpoint method

Example 4: In a similar fashion as above, one obtains for $n = 2$ and $p = 3$ the conditions

$$\begin{aligned} \gamma_0 + \gamma_1 + \gamma_2 &= 1 \\ \alpha_1\gamma_1 + \alpha_2\gamma_2 &= \tfrac{1}{2} \\ \alpha_1^2\gamma_1 + \alpha_2^2\gamma_2 &= \tfrac{1}{3} \\ \alpha_1\beta_{2,1}\gamma_2 &= \tfrac{1}{6}. \end{aligned}$$

The scheme below shows a specific solution which was also given by Heun.

$$\begin{array}{c|ccc} 0 & 0 & & \\ \frac{1}{3} & \frac{1}{3} & 0 & \\ \frac{2}{3} & 0 & \frac{2}{3} & 0 \\ \hline & \frac{1}{4} & 0 & \frac{3}{4} \end{array}$$

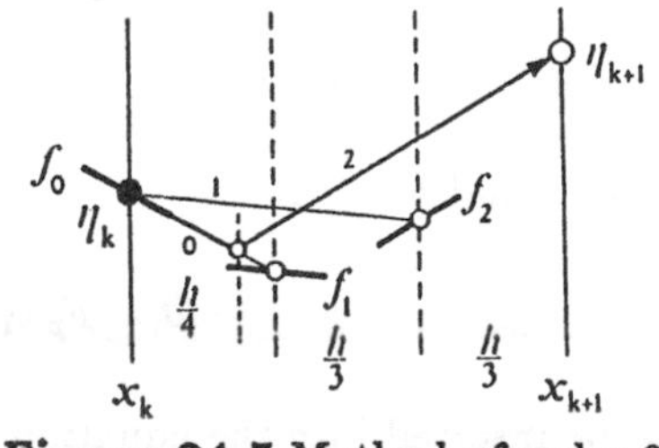

Figure 24.7 Method of order 3

Example 5: The conditions become exceedingly complex for higher orders p. As a last method the well-known method by Runge-Kutta of order $p = 4$ with $n = 3$ is given below.

0	0			
$\frac{1}{2}$	$\frac{1}{2}$	0		
$\frac{1}{2}$	0	$\frac{1}{2}$	0	
1	0	0	1	0
	$\frac{1}{6}$	$\frac{1}{3}$	$\frac{1}{3}$	$\frac{1}{6}$

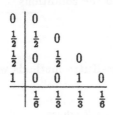

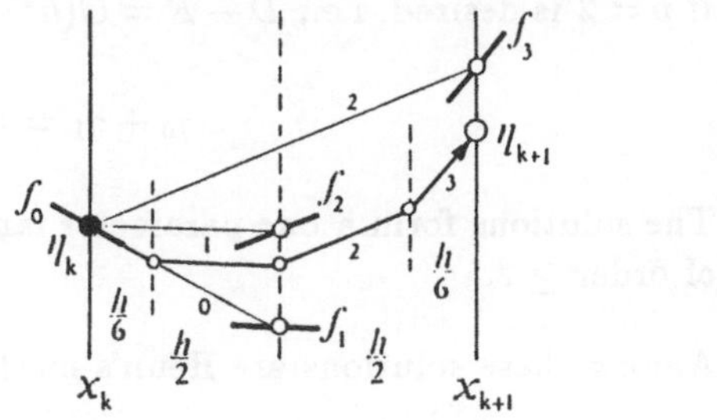

Figure 24.8 Runge-Kutta method of order 4

24.4 Error Control

The global discretization error is, as was seen in the proof of Theorem 1, just an accumulation of the local discretization errors times the associated step size if the influence of the Lipschitz constant L is neglected. The local discretization error can therefore be used to estimate the global error.

However, if a constant step size is chosen while the local discretization error varies a lot, it is difficult to predict the error growth. If on the other hand, the step size is adaptively chosen so as to keep the local error approximately constant, the global error is roughly proportional to this constant.

Consequently, one would like to get a grasp on the local discretization error which lends itself to a simple computation. For this task, two one-step methods of orders p and $p+1$ can be used.

Let F_p and F_{p+1} be the respective increment functions. If F_p and f satisfy certain regularity conditions, one can derive for the local discretization error the asymptotic expansion

$$l_k(h) = F_p - D = c_p(x_k)h^p + O(h^{p+1}).$$

Its verification is skipped here.

On the other hand one has

$$D = F_{p+1} + O(h^{p+1})$$

which can be inserted into the equation above,

$$l_k = F_p - F_{p+1} + O(h^{p+1}).$$

On dropping the last term, a simple approximation for l_k is obtained which can be used to control the step size h (or actually l_k).

Suppose the allowable error ε is attained for some step size h_ε. Furthermore, let $e := F_p - F_{p-1}$ be an approximation to l_k for some initial step size h. Together one has

$$\varepsilon = l_k(h_\varepsilon) \approx c_p h_\varepsilon^p \quad \text{and} \quad e \approx l_k(h) \approx c_p h^p \,.$$

This allows to eliminate c_p whence

$$h_\varepsilon = h \sqrt[p]{\left|\frac{\varepsilon}{e}\right|} \,.$$

These ideas lead to the following algorithm:

One-Step Method with Step Size Control

Input: $y' = f(x, y)$; a, b; y_0; h; tolerance ε; F_p, F_{p+1} increment rules
Output: $\eta := y(b)$

1 Set $x := a$, $\eta := y_0$.
2 If $x + h \geq b$, then $h := b - x$.
3 If $x \geq b$, then **stop**.
4 Determine $F_p(x, \eta, h, f)$, $F_{p+1}(x, \eta, h, f)$.
5 $e := |F_p - F_{p+1}|$,
6 $h_\varepsilon := h \sqrt[p]{\frac{\varepsilon}{e}}$.
7 If $h_\varepsilon < h$, then $h := h_\varepsilon$, go to **4**.
8 Set $x := x + h$, $\eta := \eta + hF_{p+1}$, $h := h_\varepsilon$,
9 go to **2**.

Remark 1: For a system $\mathbf{y}' = \mathbf{f}(x, \mathbf{y})$ of n first-order differential equations one has vectors $\boldsymbol{\eta}$, $\mathbf{F}$, and also $\mathbf{e}$. Line **5** is then replaced by $e := \|\mathbf{F}_p - \mathbf{F}_{p+1}\|_\infty$. The most inaccurate coordinate governs the step size.

Remark 2: It is also possible to derive an approximation e by comparing a step of size h with two steps of size $\frac{h}{2}$. However, this method is far more expensive than the one presented.

Sarafyan, England and Fehlberg have given pairs of Runge-Kutta methods of orders p and $p+1$ with matching coefficients $\beta_{i,j}$. This cuts down the amount of computation in the step size control algorithm. The following example is by Fehlberg.

Example 6: For $n = 1$, $p = 2$ and $n = 2$, $p = 3$ the coefficients are

$$\begin{array}{c|cc} 0 & 0 & \\ 1 & 1 & 0 \\ \hline & \frac{1}{2} & \frac{1}{2} \end{array} \qquad \begin{array}{c|ccc} 0 & 0 & & \\ 1 & 1 & 1 & \\ \frac{1}{2} & \frac{1}{4} & \frac{1}{4} & 0 \\ \hline & \frac{1}{6} & \frac{1}{6} & \frac{2}{3} \end{array}$$

The method of order 2 was considered in Example 2.

24.5 Notes and Exercises

1. Every convergent one-step method for which the increment function F is continuous and satisfies a Lipschitz condition has a local discretization error of order at least one.
2. Derive the equations of Example 4.
3. The initial value problem $y' = -\lambda y$, $y(0) = 1$, $\lambda > 0$, has the solution $y = e^{-\lambda x}$. However, the solutions generated by the explicit Euler method

$$\eta_{k+1} := \eta_k + hf(x_k, \eta_k)$$

become arbitrarily small for large x only if $|1 - h\lambda| < 1$. On the other hand, the solutions generated by the **implicit Euler method**

$$\eta_{k+1} := \eta_k + hf(x_k, \eta_{k+1})$$

converge to zero if x becomes large for every h.

4. In practical applications the new step size is computed as

$$h_\varepsilon := 0.8 h_0 \sqrt[p]{\frac{\varepsilon}{|e| + \delta}} \quad \text{where } \delta > 0$$

in order to avoid to many repetitions of lines **4**, **5**, **6**, and **7** in the algorithm. The growth of the step size can be limited to a factor of 10 with $\delta := \varepsilon(0.08)^p$. This is important for small $|e|$.

25 Linear Multi-Step Methods for Differential Equations

Besides the one-step methods there are the multi-step methods to solve first-order differential equations. These methods determine η_{k+n} from several previous values $\eta_k, \ldots, \eta_{k+n-1}$. Therefore, an approximate solution has to be known initially at n equidistant mesh points.

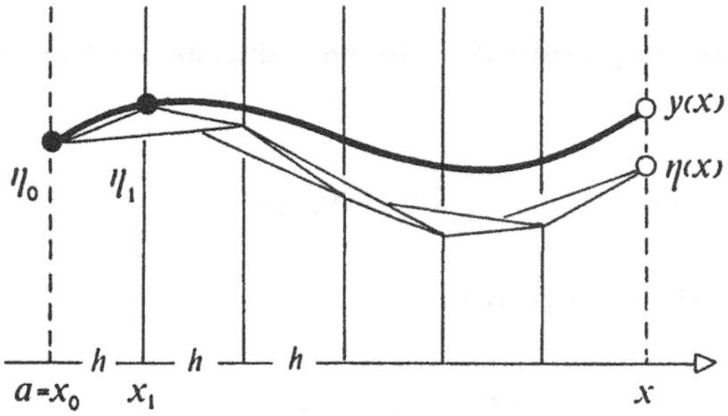

Figure 25.1 Multi-step method

25.1 Discretization

One way of discretizing the differential equation $y' = f(x, y)$ over an equidistant mesh $x_k := a + k \cdot h$ with step size $h := \frac{x-a}{m}$ is to replace the differential quotient by a difference quotient from the table in **22.1**.

For instance, the formula

$$y'_k \approx \frac{1}{2h}(-3y_k + 4y_{k+1} - y_{k+2})$$

for the values η_k of an approximate solution to the differential equation $y' = f(x, y)$ gives rise to the recursion

(1) $$\eta_{k+2} - 4\eta_{k+1} + 3\eta_k = -2hf_k\,, \quad f_k := f(x_k, \eta_k)\,.$$

The formula

$$y'_{k+1} \approx \frac{1}{2h}(y_{k+2} - y_k)$$

gives rise to the **midpoint rule**

$$\eta_{k+2} - \eta_k = 2hf_{k+1}\,.$$

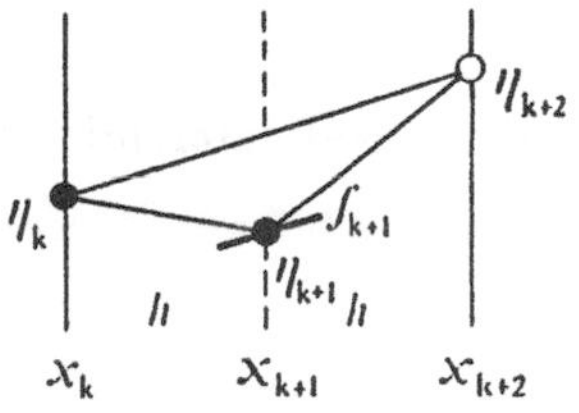

Figure 25.2 Midpoint rule

Another possibility is to integrate the differential equation $y' = f(x, y)$ formally

$$y(x) = y_k + \int_{x_k}^{x} f(t, y(t))dt$$

and to carry out the integration by one of the formulas in **22.3**.

For example, on using the **trapezoidal rule** one obtains for the values η_k the recursion

(2) $$\eta_{k+1} = \eta_k + \frac{h}{2}(f_k + f_{k+1})$$

and from Simpson's rule **Milne's method**

(3) $$\eta_{k+2} = \eta_k + \frac{h}{3}(f_k + 4f_{k+1} + f_{k+2}).$$

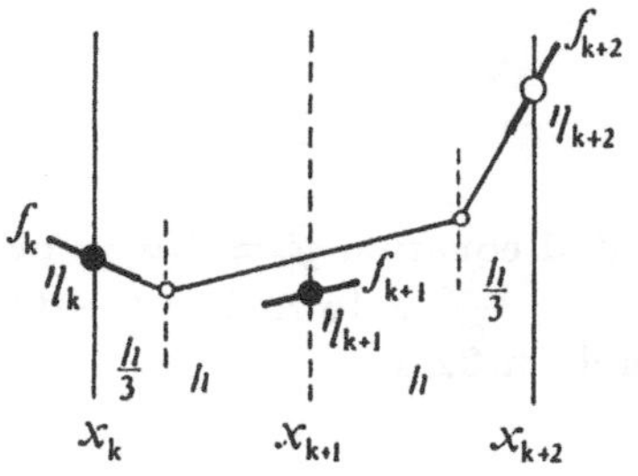

Figure 25.3
Milne's method

In general, the recursion of a **linear multi-step method** has the form

(4) $$\eta_{k+n} + \alpha_{n-1}\eta_{k+n-1} + \cdots + \alpha_0\eta_k = h(\beta_n f_{k+n} + \cdots + \beta_0 f_k)$$

with suitable constants α_p, β_r.

It is called **n-step method** if $|\alpha_0| + |\beta_0| \neq 0$, and is called **explicit** if $\beta_n = 0$ because then the $f(x_{k+r}, \eta_{k+r})$ at the right side involve only known η_{k+r}. The midpoint method is explicit for instance.

On the other hand, the method is said to be **implicit** if $\beta_n \neq 0$. Milne's method, for example, is implicit.

Not all methods which can be derived from numerical differentiation or integration are useful, though.

25.2 Convergence of Multi-Step Methods

The convergence behavior of a multi-step method is also affected by the n **starting values** $\eta_0, \eta_1, \ldots, \eta_{n-1}$. It is certainly reasonable to require that these values converge to the exact solution for a refinement of the mesh $x_k = a + kh$ with $h = \frac{x-a}{m}$, x fixed and $m \to \infty$.

Therefore, a linear multi-step method is said to be **convergent** if for each Lipschitz continuous initial value problem $y' = f(x, y)$ with $y_0 = y(a)$ the approximation η_m calculated for $y(x)$ converges to $y(x)$ as $m \to \infty$ whenever the starting values $\eta_0, \ldots, \eta_{n-1}$ converge to y_0.

The following example shows that even methods which can be deemed reasonable do not need to converge.

Example 1: Consider the initial value problem $y' = 0$ with $y(0) = 0$ for which $y(1)$ is to be found. Using (1) and the deviating starting values $\eta_0 := 0$, $\eta_1 := \frac{\varepsilon}{m}$ for $\varepsilon > 0$ one gets

$$\eta_{k+2} - 4\eta_{k+1} + 3\eta_k = 0$$

since $f = 0$. Induction gives

$$\eta_m = \frac{1}{2}\frac{\varepsilon}{m}(3^m - 1).$$

This value tends to infinity as $m \to \infty$ but not to $y(1) = 0$. Hence, the method (1) is not **convergent**.

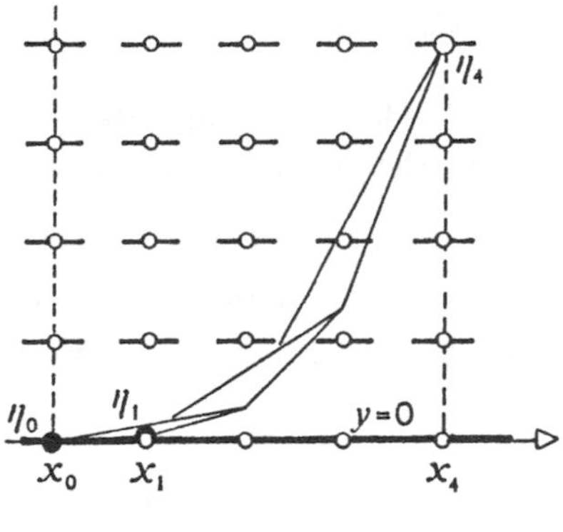

Figure 25.4
Test problem

25.3 Root Condition

The initial value problem

$$y'(x) = 0, \qquad y(0) = 0$$

together with the starting values

$$\eta_0 := \cdots := \eta_{n-2} := 0\,, \qquad \eta_{n-1} := \frac{\varepsilon}{m}\,, \quad \varepsilon > 0\,,$$

is used as a general **test problem** for arbitrary multi-step methods to find necessary conditions for convergence.

From rule (4) one obtains for the test problem the difference equation

$$\eta_{k+n} + \alpha_{n-1}\eta_{k+n-1} + \cdots + \alpha_0\eta_k = 0\,.$$

Here, one is interested in the solution to the inaccurate starting values

$$\eta_0 = \cdots = \eta_{n-2} = 0\,, \qquad \eta_{n-1} = \frac{\varepsilon}{m}\,.$$

The difference equation has the characteristic polynomial

$$\sigma(\lambda) := \lambda^n + \alpha_{n-1}\lambda^{n-1} + \cdots + \alpha_0\,.$$

One can now elaborate on the proof given in **16.3** for the convergence of Bernoulli's method and gets:

Theorem 1: $\lim_{m\to\infty} \eta_m = 0$ *holds if and only if all zeros of* $\sigma(\lambda)$ *with absolute value* 1 *are simple and all other zeros are absolutely smaller than* 1.

This condition on $\sigma(\lambda)$ is known as **root condition**. The proof is omitted here. Since $\lim_{m\to\infty} \eta_m$ is the limit of the solutions calculated by any multi-step method at $x = 1$, it is necessary for convergent methods to fulfill the root condition.

Example 2: The method of Example 1 has the characteristic polynomial $\sigma(\lambda) = \lambda^2 - 4\lambda + 3$ of which 1 and 3 are the roots. Hence, this method is divergent as already mentioned.

Example 3: Both Milne's method and the midpoint method have the characteristic polynomial $\sigma(\lambda) = \lambda^2 - 1$ with the distinct roots ± 1. Both methods converge for the test problem.

25.4 Sufficient Convergence Conditions

Every linear multi-step method is by virtue of (4) associated with a difference operator $\mathcal{L}$, namely $\mathcal{L}$ is defined for any function $g(x)$ which is continuously differentiable in [0,1] as follows

$$\mathcal{L}(g_k, h) := g_{k+n} + \alpha_{n-1}g_{k+n-1} + \cdots + \alpha_0 g_k - h(\beta_n g'_{k+n} + \cdots + \beta_0 g'_k)$$

where

$$g_{k+r} := g(x_k + rh)\,, \qquad g'_{k+r} := g'(x_k + rh)\,.$$

Let g be a solution of the equation $y' = f(x, y)$. Then, because of (4)

$$\mathcal{L}(g_k, h) = g_{k+n} - \eta_{k+n} + h\beta_n \left(f(x_{k+n}, \eta_{k+n}) - g'_{k+n}\right)$$

where η_{k+n} is the approximation to g'_{k+n} generated by the multiple-step method from the values g_{k+r} and g'_{k+r}, $r = 0, \ldots, n-1$.

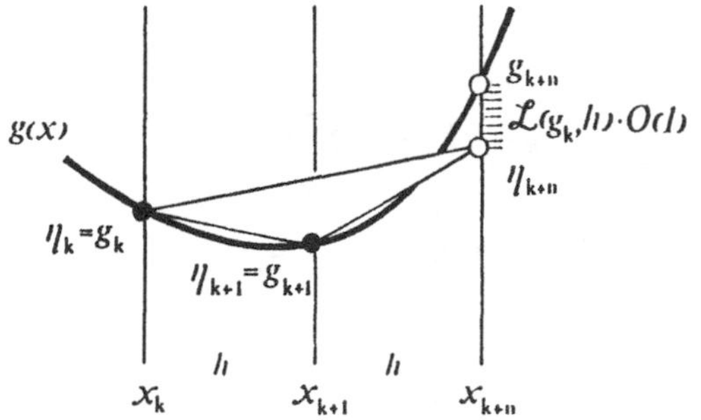

Figure 25.5
Difference operator

Because f satisfies a Lipschitz condition, one gets

$$\begin{aligned}\left|f(x_{k+n}, \eta_{k+n}) - g'_{k+n}\right| &= \left|f(x_{k+n}, \eta_{k+n}) - f(x_{k+n}, g_{k+n})\right| \\ &\le L\left|g_{k+n} - \eta_{k+n}\right|\end{aligned}$$

and therefore

$$\mathcal{L}(g_k, h) \ge \left|g_{k+n} - \eta_{k+n}\right| (1 - h\,|\beta_n|\,L)\,.$$

Consequently, $|\mathcal{L}(g_k, h)|$ should be small for arbitrary g_k in order to consider the multi-step method useful. One says:

A linear multi-step method is of **order p** if

$$\mathcal{L}(g_k, h) = (g_{k+n} - \eta_{k+n})O(1) = O(h^{p+1})$$

for all functions $g(x)$ which are continuously differentiable in $[a, b]$.

This entails that $\mathcal{L}(\mathrm{pol}(x_k), h) = 0$ for all polynomials $\mathrm{pol}(x)$ up to degree p since $\mathcal{L}(\mathrm{pol}(x_k), h)$ is a polynomial in h of degree $\le p$. One can even show the converse by a Taylor expansion of $\mathcal{L}(g_k, h)$ for sufficiently often differentiable g. The following theorem is a consequence of this fact.

Theorem 2: *A linear multi-step method is of order* 1 *if*

$$1 + \alpha_{n-1} + \cdots + \alpha_0 = 0$$

and

$$n + (n-1)\alpha_{n-1} + \cdots + \alpha_1 = \beta_n + \cdots + \beta_0 .$$

For the proof the Taylor expansion of $g(x_k + t)$ about $t = 0$ is considered. Since the terms of order 2 do not matter and since $\mathcal{L}$ is linear in the g, it suffices to look at the cases $g = 1$ and $g = x$. So, one obtains

$$\mathcal{L}(1, h) = \sum_{r=0}^{n} \alpha_r = 0$$

where $\alpha_n := 1$ and

$$\begin{aligned}\mathcal{L}(x_k, h) &= \sum_{r=0}^{n} \alpha_r (x_k + rh) - h \sum_{r=0}^{n} \beta_r \\ &= x_k \sum_{r=0}^{n} \alpha_r + h \sum_{r=0}^{n} (r\alpha_r - \beta_r) \\ &= h \sum_{r=0}^{n} (r\alpha_r - \beta_r) = 0\end{aligned}$$

as claimed.

Example 4: For the integration via the trapezoidal rule, $\mathcal{L}$ takes on the form

$$\mathcal{L}(g_k, h) = g_{k+1} - g_k - \frac{h}{2}(g'_{k+1} + g'_k) .$$

Hence, $\alpha_1 = 1$, $\alpha_0 = -1$, $\beta_1 = \frac{1}{2}$, $\beta_0 = \frac{1}{2}$ and

$$\begin{aligned}\mathcal{L}(1, h) &= 1 - 1 = 0 , \\ \mathcal{L}(x_k, h) &= (x_k + h) - x_k - \frac{h}{2} 2 = 0 , \\ \mathcal{L}(x_k^2, h) &= (x_k + h)^2 - x_k^2 - \frac{h}{2}(2(x_k + h) + 2x_k) = 0 .\end{aligned}$$

However,

$$\mathcal{L}(x_k^3, h) = (x_k + h)^3 - x_k^3 - \frac{h}{2}(3(x_k + h)^2 + 3x_k^2) = -\frac{1}{2}h^3 \neq 0 .$$

Hence, the method (2) is of order 2.

The following fundamental theorem characterizes the **convergence** of a linear multi-step method.

Theorem 3: *A linear multi-step method is convergent if and only if it satisfies the root condition and is at least of linear order.*

The proof is beyond the scope of this book.

Remark 1: In setting up a multi-step method one can try to choose the α_r and β_r in (4) so as to maximize the order p, i.e., $\mathcal{L}(\mathrm{pol}(x_k), h)$ has to vanish for all polynomials of degree $\leq p$. This gives rise to a linear system for the α_r, β_r. Additionally, the root condition has to be satisfied, see also Exercise 4.

25.5 Starting Values

The n starting values $\eta_0, \ldots, \eta_{n-1}$ for a linear n-step method are usually procured by a one-step method **24** which is applied to the same initial value problem. The order of the one-step method should be at least as big as the one of the multi-step method.

The step size for the computation of $\eta_1, \ldots, \eta_{n-1}$ should be chosen small enough to provide small enough errors, but not so small that rounding-error build-up spoils the theoretically achievable accuracy. Distorted starting values might affect the accuracy of further computations. The initial step size can be chosen by the methods of **24.4**.

25.6 Predictor-Corrector Methods

Implicit multi-step methods are superior to explicit methods because the same accuracy is achieved with a larger step size.

In each step of such a method a non-linear equation (4) of the form

$$\eta_{k+n} = h \sum_{r=0}^{n} \beta_r \, f_{k+r} - \sum_{r=0}^{n-1} \alpha_r \, \eta_{k+r} =: P(\eta_{k+n})$$

is to be solved where f_{k+n} depends on η_{k+n}. This is done iteratively using pairs of linear multi-step methods.

An initial approximation $\eta_{k+n}^{(0)}$ is determined by an explicit linear multi-step method, the **predictor**. This value is improved by the iteration rule, the so called **corrector**,

(5) $$\eta_{k+n}^{(i+1)} := P(\eta_{k+n}^{(i)})$$

of the implicit method.

The order of the explicit predictor should be at most 1 less than the order of the implicit corrector. In this case one iteration step is usually sufficient.

Remark 2: A sufficient condition for the convergence of the iteration (5) is provided by the fixed-point theorem in **13.1**, namely P must be contractive. Since

$$|P(\eta) - P(\zeta)| = |h\beta_n\,(f(x_k,\eta) - f(x_k,\zeta))| \le h\,|\beta_n|\,L\,|\eta - \zeta|,$$

this is the case for sufficiently small h.

Example 5: On using the rule $F = hf_k$ and the trapezoidal rule one obtains the

predictor $\eta_{k+1}^{(0)} := \eta_k + hf_k$ of order $p = 1$ and the

corrector $\eta_{k+1}^{(1)} := \eta_k + \frac{h}{2}(f_k + f_{k+1})$ of order $p = 2$.

The corresponding method is the method of Heun in Example 2 of **24.2**.

Example 6: On using the rule $F = \frac{h}{12}(5f_k - 16f_{k+1} + 23f_{k+2})$ for an overhanging interval in **22.3** and Simpson's rule one obtains the

predictor $\eta_{k+3}^{(0)} := \eta_{k+2} + \frac{h}{12}(5f_k - 16f_{k+1} + 23f_{k+2})$ of order $p = 3$,

and Milne's rule for the

corrector $\eta_{k+3}^{(1)} := \eta_{k+1} + \frac{h}{3}(f_{k+1} + 4f_{k+2} + f_{k+3})$ of order $p = 4$.

Remark 3: In practice, one prefers methods of higher order, though.

25.7 Step Size Control

An approximation of the discretization error after one step is also used in multistep methods to proportion the step size. Under certain regularity assumptions, the error $e = \mathcal{L}(y_k, h)$ has the asymptotic expansion

$$\mathcal{L}(y_k, h) = c_{p+1}h^{p+1} + O(h^{p+2}).$$

As in **24.4** one can determine an approximation for $\mathcal{L}$ from the difference between the predictor and corrector at x_{k+n}.

Consider for example the predictor $\eta_{k+3}^{(0)} := \eta_{k-1} + \frac{4h}{3}(2f_k - f_{k+1} + 2f_{k+2})$ and the corrector of Example 6 (Milne's method). The corresponding error expansions are

$$y_{k+3} - \eta_{k+3}^{(0)} = c_5^{(0)}h^5 + O(h^6),$$

and

$$y_{k+3} - \eta_{k+3}^{(1)} = c_5^{(1)}h^5 + O(h^6),$$

where $c_5^{(0)} = -28c_5^{(1)}$. On eliminating the constants $c_5^{(0)}$, $c_5^{(1)}$ one gets

$$g_{k+3} - \eta_{k+3}^{(1)} = -\frac{1}{29}(\eta_{k+3}^{(1)} - \eta_{k+3}^{(0)}) + O(h^6).$$

Thus

$$e = -\frac{1}{29}\frac{1}{h}(\eta_{k+3}^{(1)} - \eta_{k+3}^{(0)})$$

is a good approximation for the local discretization error.

Suppose that one wishes to keep the local discretization error within some tolerance $\varepsilon > 0$. Then one can control the step size as in a one-step method, i.e., the improved step size is

$$h_\varepsilon := h\sqrt[4]{\left|\frac{\varepsilon}{e}\right|}.$$

Example 7: The one-step method of order 4 in **24.3** is a good choice when computing the starting values η_1, η_2 for the multi-step method of Example 6. The step size h stays fixed then. Secondly, one determines f_0, f_1, f_2 through the differential equation. With this, the multi-step method can be started. At the point $x_3 := a + 3h$ one calculates successively

$$\eta_3^{(0)}, \quad f_3^{(0)} := f(x_3, \eta_3^{(0)}), \quad \eta_3^{(1)} \quad \text{and} \quad f_3 := f(x_3, \eta_3^{(1)}),$$

then analogously at $x_4 := a + 4h$ the values $\eta_4^{(0)}, \ldots, f_4$, etc.

If eventually $x_{k+2} + h > b$, then one determines $\eta(b)$ using the one-step method again with the starting values x_{k+2}, η_{k+2} and step size $h = b - x_{k+2}$.

Controlling the step size is more involved than for one-step methods in **24.4**. Utilizing previously computed values η_k and f_k is only possible if the step size is **halved** or **doubled**.

Halving: Before calculating η_{k+3} by the multi-step method one sets

$$x_k := x_{k+1}, \quad \eta_k := \eta_{k+1}, \quad f_k := f_{k+1},$$

and $h := \frac{h}{2}$. Then one determines η_{k+1} by the one-step method, subsequently f_{k+1} by means of the differential equation, and finally η_{k+3} by the multi-step method, etc. It is possible to repeat the halving successively.

Doubling: Before calculating η_{k+3} by the multi-step method one sets

$$\begin{aligned} x_{k+1} &:= x_k, \quad & \eta_{k+1} &:= \eta_k, \quad & f_{k+1} &:= f_k, \\ x_k &:= x_{k-2}, \quad & \eta_k &:= \eta_{k-2}, \quad & f_k &:= f_{k-2}, \end{aligned}$$

and $h = 2h$. Then one determines η_{k+3}, etc. To make sure that all values needed for the computation are available two ordinary steps with constant step size should be performed before each doubling.

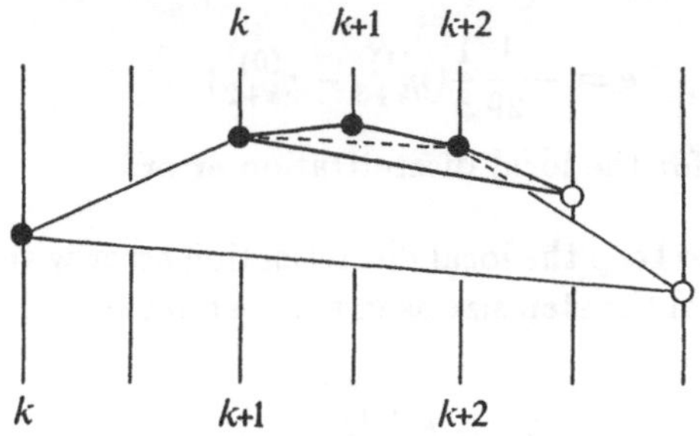

Figure 25.6
Step size control

25.8 Comparing One- and Multi-Step Methods

Although the algorithm in **25.7** is far from optimal, it displays the essential drawback of multi-step methods. They are cumbersome with regard to step size control. Only a doubling is useful to increase the step size and for a reduction intermediate values have to be obtained from a one-step method.

However, if the function flows fairly evenly, only a few changes of the step size will be necessary. Then preference should be given to multi-step methods, since $y' = f(x, y)$ has to be evaluated only twice for each step.

25.9 Notes and Exercises

1. The polynomial $\sigma(\lambda)$ of the methods which are derived by numerical integration over the interval $[x_{n-q}, x_n]$ has the form

$$\sigma(\lambda) = \lambda^n - \lambda^{n-q}.$$

Hence these methods satisfy the root condition.

2. Milne's method is of order 4.

3. In analogy to **25.4** the requirement $\mathcal{L}(x_k^s, h) = 0$ for a multi-step method of order p translates into

$$\sum_{r=0}^{n}(r^s\alpha_r - sr^{s-1}\beta_r) = 0, \quad s = 0, 1, \ldots, p.$$

These are $p + 1$ linear conditions for the α_r, β_r.

4. A differential equation is called **stable** if the difference of any two solutions for sufficiently close initial values approaches zero as $x \to \infty$.
5. The prototype of a stable differential equation is

$$y' = -\lambda y \text{ for } \lambda > 0 .$$

6. Some linear multi-step methods are unsuitable for the integration of stable differential equations since the numerical solution begins to oscillate. This is the case with Milne's rule, for instance.

26 The Methods by Ritz and Galerkin

Many physical and technical problems are described by variational (properties or) equations. The methods by Ritz and Galerkin are of great importance for a practical solution of these problems.

26.1 The Principle of Minimal Energy

Often, in physical problems, a function u has to be determined which minimizes the **potential energy** of some system subject to certain boundary conditions. In general, the energy is described by a functional

$$\mathcal{I}[u] = \int_G \left(\frac{1}{2} q(u,u) - l(u) \right)$$

where q is quadratic and l linear in u and its derivatives; i.e., on combining u and its ordinary or respectively partial derivatives to one column,

$$\boldsymbol{u} = [u,\, u',\, u'']^t$$

or

$$\boldsymbol{u} = [u,\, u_x,\, u_y,\, u_{xy}]^t , \quad \text{etc.,}$$

one can write $q(u,u) = \boldsymbol{u}^t Q \boldsymbol{u}$ with a symmetric matrix Q and $l(u) = \boldsymbol{u}^t \boldsymbol{l}$ with a column $\boldsymbol{l}$.

The domain of integration G denotes the region of the physical system.

Example 1: After some simplifications, the energy of a rope which is fastened at the points $x = 0$ and $x = l$ and loaded with $f = f(x)$ is described by

$$\mathcal{I}[u] = \frac{1}{2} \int_0^l (u')^2 dx - \int_0^l f \cdot u \; dx .$$

The curve of minimal energy $u(x)$ is shown in Figure 26.1.

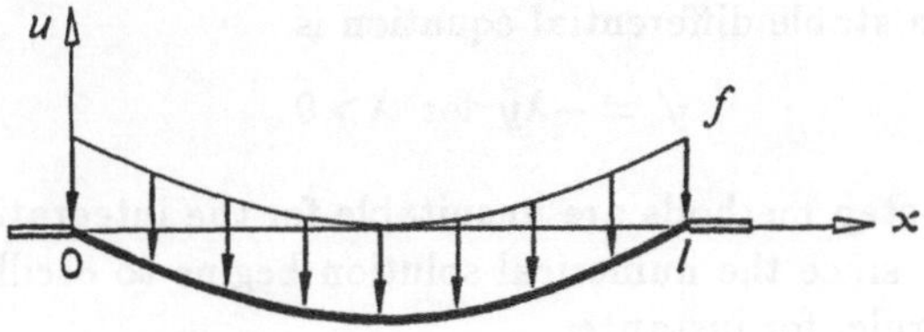

Figure 26.1 Loaded rope

Example 2: On similar simplifications, the energy of a diaphragm which is fastened at two parallel edges $x = 0$ and $x = l$ and loaded with $f = f(x, y)$ is described by

$$\mathcal{I}[u] = \frac{1}{2}\int_0^l\int_0^b (u_x^2 + u_y^2)dydx - \int_0^l\int_0^b f \cdot u \; dydx\,.$$

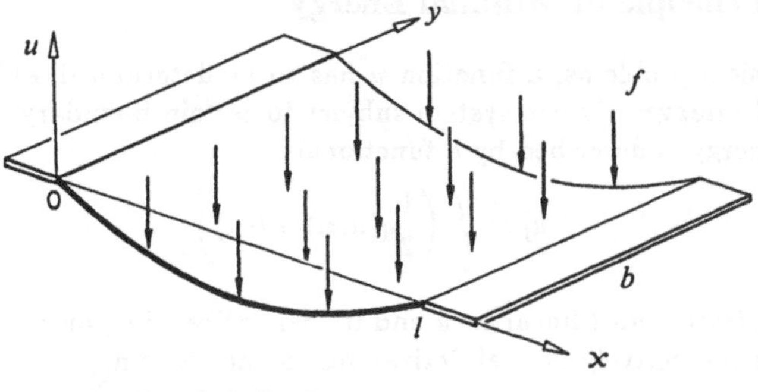

Figure 26.2 Loaded diaphragm

26.2 The Ritz Method

The energy problem is solved numerically by minimizing the functional over all linear combinations

$$v = c_1v_1 + c_2v_2 + \cdots + c_nv_n$$

of linearly independent **trial functions** v_i which satisfy the physical and mechanical requirements, i.e., which are **admissible**. Then the integral takes on the form

$$\mathcal{I}[v] = \sum_{i,j} c_ic_j\frac{1}{2}\int_G q(v_i, v_j) - \sum_i c_i\int_G l(v_i)$$

or in matrix notation

$$\mathcal{I}[v] = \frac{1}{2}[c_1, \ldots, c_n] \begin{bmatrix} \int_G q(v_1, v_1) & \cdots & \int_G q(v_1, v_n) \\ \vdots & & \vdots \\ \int_G q(v_n, v_1) & \cdots & \int_G q(v_n, v_n) \end{bmatrix} \begin{bmatrix} c_1 \\ \vdots \\ c_n \end{bmatrix} - [c_1, \ldots, c_n] \begin{bmatrix} \int_G l(v_1) \\ \vdots \\ \int_G l(v_n) \end{bmatrix}.$$

This is a quadratic form in the coefficients c_i

$$\mathcal{I}(\mathbf{c}) = \frac{1}{2}\mathbf{c}^t A\mathbf{c} - \mathbf{c}^t \boldsymbol{a}.$$

Since the term $\frac{1}{2}\mathbf{c}^t A\mathbf{c}$ corresponds to an energy, A is symmetric and also positive definite, whenever the trial functions v_i satisfy the boundary conditions. Hence, according to **8.3**, the quadratic form $\mathcal{I}(\mathbf{c})$ assumes its minimum at the solution of the linear system

$$A\mathbf{c} - \boldsymbol{a} = \mathbf{o}.$$

These equations are known as **Ritz equations**. Their solution determines the so-called **Ritz approximation** v for the exact solution u.

Example 3: On using the linear combination

$$v = c_1 v_1 + c_2 v_2 + \cdots + c_n v_n$$

for the rope in Example 1 one obtains the coefficients

$$a_{i,j} = \int_G q(v_i, v_j) = \int_0^l v_i' v_j' \; dx,$$

$$a_i = \int_G l(v_i) = \int_0^l f \cdot v_i \; dx$$

for the Ritz equations. The simplest choice for admissible trial functions are piecewise linear functions where $v_i(0) = v_i(l) = 0$.

Example 4: On using the linear combination

$$v = c_1 v_1 + c_2 v_2 + \cdots + c_n v_n$$

for the diaphragm of Example 2 one obtains the coefficients

$$a_{i,j} = \int_0^l \int_0^b \left(\frac{\partial v_i}{\partial x} \frac{\partial v_j}{\partial x} + \frac{\partial v_i}{\partial y} \frac{\partial v_j}{\partial y} \right) dy dx,$$

$$a_i = \int_0^l \int_0^b f \cdot v_i \; dy dx$$

for the Ritz equations.

26.3 Galerkin's Method

Problems of mechanics lead frequently to differential equations

$$\mathcal{D}[u] = f$$

over some domain with certain boundary conditions. $\mathcal{D}$ can be an arbitrary differential operator. The approximation of u by a linear combination

$$v = c_1 v_1 + \cdots + c_n v_n$$

of n linear independent trial functions gives a residual,

$$\varrho = \mathcal{D}[v] - f\,.$$

Following an idea of Galerkin one determines the coefficients c_i such that the means of the residual ϱ weighted by the v_i vanish, i.e., such that

$$\int_G \varrho \cdot v_i = \int_G (\mathcal{D}[c_1 v_1 + \cdots + c_n v_n] - f) v_i = 0\,, \quad i = 1, \ldots, n\,.$$

These are the **Galerkin equations** for the c_j (see also **26.5**).

Remark 1: The Galerkin equations are in general not linear. However, in the case of a linear differential operator $\mathcal{L}$ they have the form

$$\sum_j c_j \int_G \mathcal{L}[v_j] v_i - \int_G f \cdot v_i = 0\,, \quad i = 1, \ldots, n\,,$$

or in matrix notation

$$\begin{bmatrix} \int_G \mathcal{L}[v_1] v_1 & \cdots & \int_G \mathcal{L}[v_n] v_1 \\ \vdots & & \vdots \\ \int_G \mathcal{L}[v_1] v_n & \cdots & \int_G \mathcal{L}[v_n] v_n \end{bmatrix} \begin{bmatrix} c_1 \\ \vdots \\ c_n \end{bmatrix} = \begin{bmatrix} \int_G f \cdot v_1 \\ \vdots \\ \int_G f \cdot v_n \end{bmatrix}.$$

Moreover, if $\mathcal{L}$ is both

1. symmetric: $\int_G \mathcal{L}[u] v = \int_G \mathcal{L}[v] u$ for all u, v,
2. and positive definite: $\int_G \mathcal{L}[u] u > 0$ for all $u \neq 0$,

in G (considering the boundary conditions), then the matrix $A = [a_{i,j}] = \left[\int_G \mathcal{L}[v_j]v_i\right]$ is also symmetric and positive definite.

In this case the corresponding solution $v = c_1v_1 + \cdots + c_nv_n$ minimizes the integral

$$\mathcal{I}[v] = \frac{1}{2}\int_G \mathcal{L}[v]v - \int_G f \cdot v$$

as can be seen by a comparison with the Ritz equations.

Example 5: The deflection $u(x)$ of the loaded rope in Example 1 which is fastened at its ends $x = 0$ and $x = l$ satisfies the linear differential equation

$$\mathcal{L}[u] = -u'' = f \text{ with } u(0) = u(l) = 0\,.$$

Approximating the solution by a linear combination

$$v = c_1v_1 + \cdots + c_nv_n$$

yields the coefficients

$$a_{i,j} = \int_0^l (-v_j''v_i)dx\,, \quad a_i = \int_0^l f \cdot v_i \; dx$$

of the Galerkin equations. These equations are identical with the Ritz equations for the integral

$$\mathcal{I}[v] = \frac{1}{2}\int_0^l (-v''v)dx - \int_0^l f \cdot v \; dx\,.$$

Example 6: The deflection of the loaded diaphragm in Example 2 satisfies the Poisson differential equation

$$\mathcal{L}[u] = -u_{xx} - u_{yy} = f\,, \quad u(0,y) = u(l,y) = 0\,.$$

One obtains for the coefficients of the Galerkin equations

$$a_{i,j} = \int_0^l\int_0^b \left(-\frac{\partial^2 v_j}{\partial x^2} - \frac{\partial^2 v_j}{\partial y^2}\right) v_i \; dydx\,, \quad a_i = \int_0^l\int_0^b f \cdot v_i \; dydx\,.$$

26.4 Relation

The relation between the methods by Ritz and Galerkin for a linear differential operator $\mathcal{L}$ and an energy integral

$$\mathcal{I}[u] = \frac{1}{2}\int_G \mathcal{L}[u]u - \int_G f \cdot u$$

is spelled out by the following

Theorem 1: ***Assuming that $\mathcal{L}$ is symmetric and positive definite in G the function v solves $\mathcal{L}[u] = f$ in G if and only if the functional $\mathcal{I}[u] = \frac{1}{2}(\int_G \mathcal{L}[u]u) - \int_G f \cdot u$ attains its minimum at v.***

The proof is simple. Let g be an arbitrary function and $\varepsilon \in \mathbb{R}$. Then one has, since $\mathcal{L}$ is symmetric,

$$\begin{aligned}\mathcal{I}[v+\varepsilon g] &= \int_G \left(\frac{1}{2}\mathcal{L}[v+\varepsilon g] - f\right)(v+\varepsilon g) \\ &= \mathcal{I}[v] + \varepsilon \int_G (\mathcal{L}[v]-f)g + \frac{\varepsilon^2}{2}\int_G \mathcal{L}[g]g\,.\end{aligned}$$

Hence, $\partial\mathcal{I}$, the so-called first variation of $\mathcal{I}$ at v becomes

$$\partial\mathcal{I} = \frac{d}{d\varepsilon}\mathcal{I}[v+\varepsilon g]\bigg|_{\varepsilon=0} = \int_G (\mathcal{L}[v]-f)g\,.$$

It vanishes for all possible g whenever $\mathcal{I}$ is minimal at v. Hence, $\mathcal{L}[v] - f = 0$.

On the other hand, if v is the solution for $\mathcal{L}[v] - f = 0$, then one has

$$\mathcal{I}[v+\varepsilon g] = \mathcal{I}[v] + \frac{\varepsilon^2}{2}\int_G \mathcal{L}[g]g \geq \mathcal{I}[v] \text{ for all } g\,,$$

since $\mathcal{L}$ is positive definite. Thus, $\mathcal{I}$ is minimal at v.

26.5 Notes and Exercises

1. One can obtain the Galerkin equations in analogy to **9.6** with the scalar product

$$(u,v) = \int_G u \cdot v$$

and the stipulation that the residual ϱ be "orthogonal" to each v_j.

2. The energy integral of Example 5 can be transformed into the integral of Example 1 by means of integration by parts and the boundary conditions.
3. Analogously, with the aid of Green's identity one can transform the energy integral associated with Example 6 to the form of Example 2.
4. As with $\boldsymbol{u}$ in **26.1** one can form vectors $\boldsymbol{v}_j$ containing v_j and its derivatives and furthermore the matrix $V = [\boldsymbol{v}_1, \ldots, \boldsymbol{v}_n]$. The matrix V is a wide matrix in general. Using it, one gets

$$A = \int_G V^t Q V \quad \text{and} \quad \boldsymbol{a} = \int_G V \boldsymbol{l}.$$

5. The Ritz method applied to problems of free dynamic leads to the generalized eigenvalue problem

$$A\boldsymbol{c} = \lambda B\boldsymbol{c}.$$

6. The Ritz method can also be used to find an approximation for the minimum of a general non-quadratic functional. The coefficients c_i are determined from the equations

$$\frac{\partial \mathcal{I}}{\partial c_i} = 0, \quad i = 1, \ldots, n.$$

In the quadratic case these are the linear Ritz equations.

27 The Finite Element Method

A suitable choice of the trial functions v_i is of great importance for the Ritz and Galerkin methods. For numerical reasons, finite elements are particularly suited for this purpose. They allow for a uniform description of systems with widely varying domains G.

27.1 Finite Elements

The finite element method is based on three basic ideas:

(1) The partition of the integral domain G into subdomains G_k, the so-called **elements**.

(2) A split of the trial functions v_i into segments $v_{i,k}$ with the support G_k.

(3) A selection of v_i's with as small a support as possible, i.e., v_i's which vanish on as many G_k as possible.

Property (1) leads to the determination of the coefficients $a_{i,j}$ and a_i of the Ritz equations by integration with respect to the subdomains

$$a_{i,j,k} = \int_{G_k} g(v_i, v_j), \qquad a_{i,k} = \int_{G_k} f \cdot v_i$$

and subsequent summation

$$a_{i,j} = \sum_k a_{i,j,k}, \qquad a_i = \sum_k a_{i,k}.$$

Property (2) permits the construction of the v_i from suitable segments $v_{i,k}$.

Many $a_{i,j,k}$ are zero as a consequence of property (3), namely those for which G_k does not contribute to the support of v_i or v_j. Hence, if the supports of v_i and v_j are disjoint, then one even has $a_{i,j} = 0$.

27.2 Univariate Splines

The simplest case is that of a one-dimensional domain G with a uniform partition. A good choice for the trial functions are the B-splines $N_i^n(x)$ of degree n. Their linear combinations

$$s(x) = \sum_i d_i \, N_i^n(x)$$

are called **splines**. The derivatives of a spline of degree n is a spline of degree $n-1$. Thus, the B-splines of degree 0 enter the picture as the derivatives of linear splines.

The **constant B-spline** $N_i^0(x)$ has the support $[i, i+1]$ and assumes the value 1 on $[i, i+1)$.

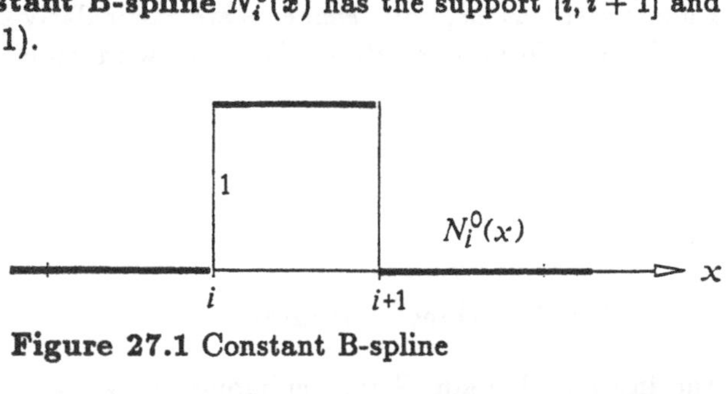

Figure 27.1 Constant B-spline

Any linear combination of these B-splines is a step function; its derivative vanishes everywhere, some precautions at the integers are needed though.

The **linear B-spline** $N_i^1(x)$ has the support $[i, i+2]$, is piecewise linear and assumes the value 1 at $i+1$. It has the Bézier ordinates $\ldots, 0, 1, 0, \ldots$.

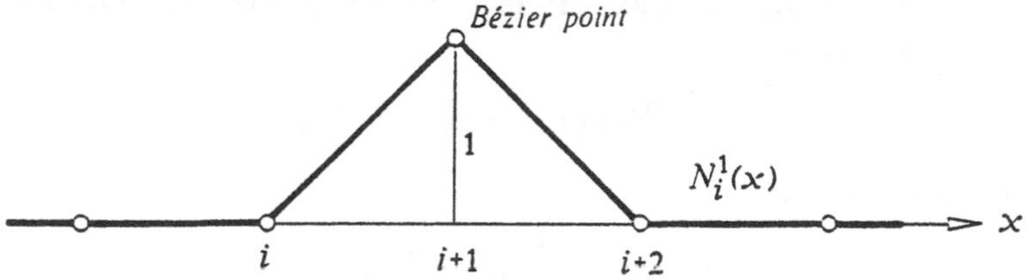

Figure 27.2 Linear B-spline

A linear combination of linear B-splines is piecewise linear and continuous. It coincides with its Bézier polygon. The first derivative of a linear B-spline is a piecewise constant spline:

$$\frac{d}{dx}N_i^1(x) = N_i^0(x) - N_{i+1}^0(x).$$

The **quadratic B-spline** $N_i^2(x)$ has the support $[i, i+3]$, is piecewise quadratic and has the Bézier ordinates $\ldots, 0, \frac{1}{2}, 1, \frac{1}{2}, 0, \ldots$.

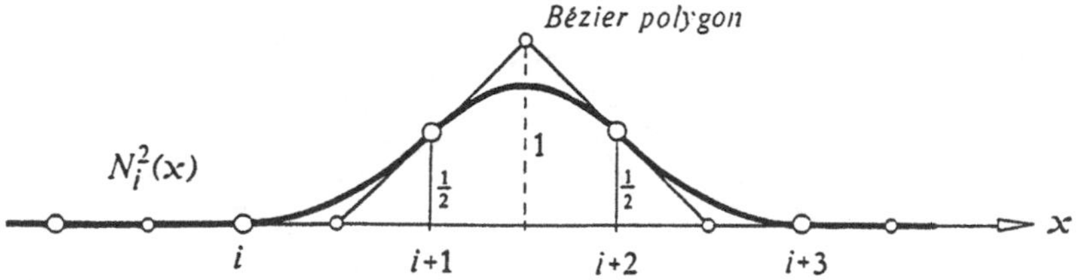

Figure 27.3 Quadratic B-spline

The linear combination is piecewise quadratic and differentiable. The first derivative of a quadratic B-spline is a linear spline:

$$\frac{d}{dx}N_i^2(x) = N_i^1(x) - N_{i+1}^1(x),$$

the second derivative a step function:

$$\frac{d^2}{dx^2}N_i^2(x) = N_i^0(x) - 2N_{i+1}^0(x) + N_{i+2}^0(x).$$

The **cubic B-spline** $N_i^3(x)$ was already introduced in **21.3** with the notation $N_{i+2}(x)$.

Remark 1: The Bézier ordinates of a spline are the linear combinations of the B-spline Bézier ordinates since linear combinations and the expansion over each segment by Bernstein polynomials (cf. **20.2**) can be exchanged.

27.3 Bivariate Splines

B-splines over rectangular partitions $x = i$, $y = j$ of the xy-plane are products of B-splines in x and in y

$$N_{i,j}^{n,m}(x,y) = N_i^n(x)\, N_j^m(y)\,.$$

The linear combinations

$$s(x,y) = \sum_{i,j} d_{i,j}\, N_{i,j}^{n,m}(x,y)$$

of these products are spline surfaces of degree n, m. Their partial derivatives are spline surfaces of lower degree, e.g.,

$$\begin{aligned}\frac{\partial}{\partial x} N_{i,j}^{n,m}(x,y) &= \frac{d}{dx} N_i^n(x) \cdot N_j^m(y) \\ &= N_{i,j}^{n-1,m}(x,y) - N_{i+1,j}^{n-1,m}(x,y)\,, \quad \text{etc.}\end{aligned}$$

Figure 27.4 shows a bilinear and Figure 27.5 a biquadratic B-spline.

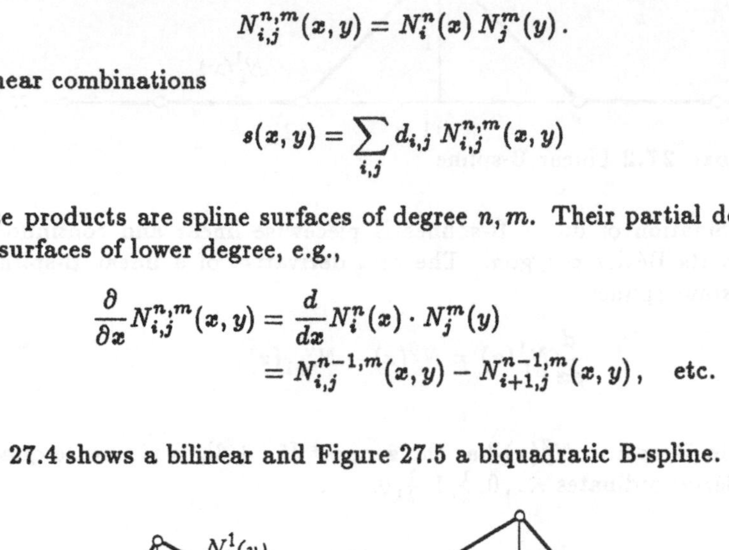

Figure 27.4 Bilinear B-spline

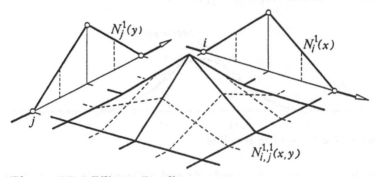

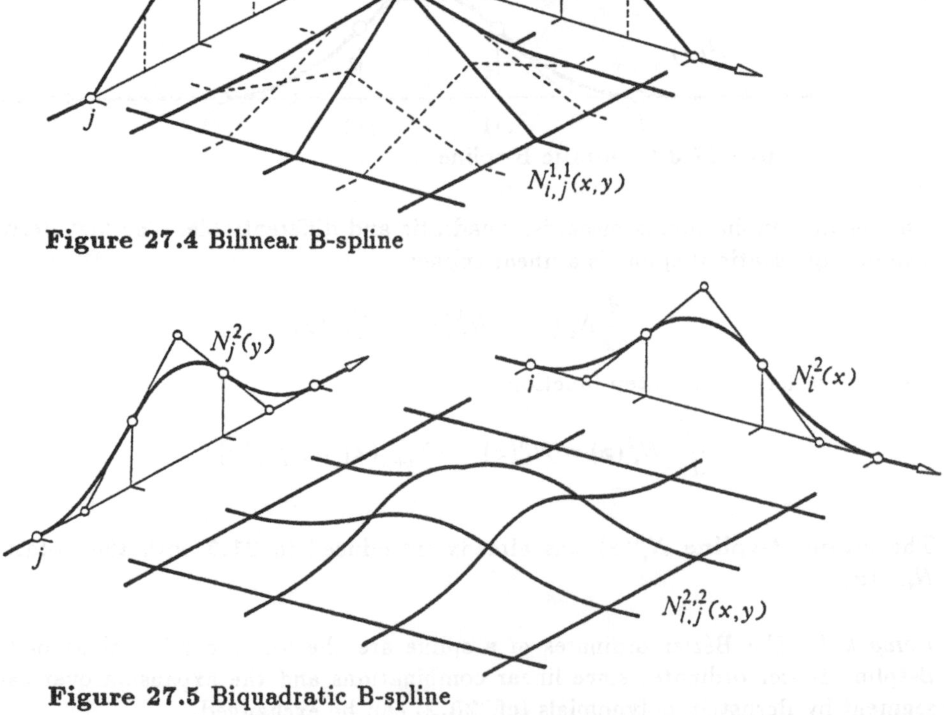

Figure 27.5 Biquadratic B-spline

Remark 2: The Bézier ordinates of such a bivariate B-spline are products of the Bézier ordinates of univariate B-splines.

Remark 3: Note, that a general bivariate spline can usually not be obtained as the product of two univariate splines, i.e., the product and the linear combination of univariate B-splines do not commute.

This fact is emphasized by the name **tensor product** given to the set of all linear combinations of the bivariate B-splines with fixed degree n, m.

Splines on irregular partitions of G are much harder to deal with. Exceptions are the piecewise constant spline and the piecewise linear spline over some triangulation of the xy-plane. A constant B-spline forms a prism as in Figure 27.6 while the linear B-spline forms a pyramid as in Figure 27.7.

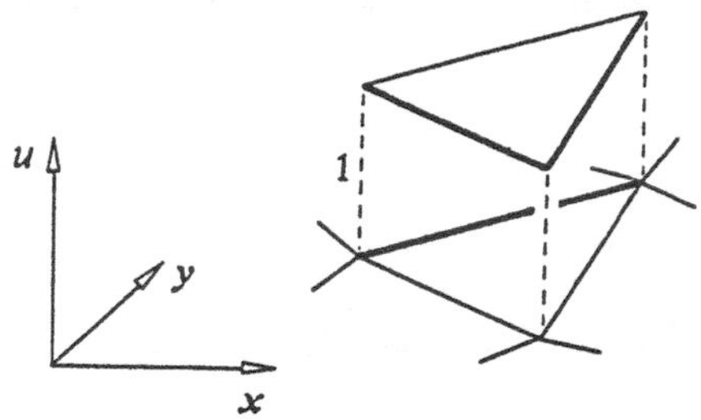

Figure 27.6 Constant B-spline over a triangulation

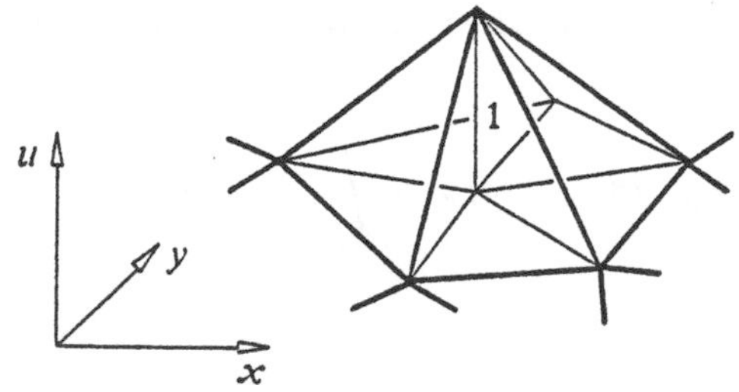

Figure 27.7 Linear B-spline over a triangulation

The first partial derivatives of such a linear B-spline are piecewise constant. It is because of their simple structure that the use of linear B-splines known as the **method of plates** has become popular in varied applications.

27.4 Numerical Examples

As practical examples, the deflection of the loaded rope and the loaded diaphragm from **26.1** are approximated by the **finite element method**.

Example 1: Let $l = 4$ be the length of the rope of **26.1** and assume that it is loaded by $f = -0.4$ on the right half from $l/2$ to l. Thus, the energy

$$\frac{1}{2}\int_0^4 (u')^2 dx - \int_0^4 f \cdot u \; dx$$

is to be minimized approximately.

The trial function

$$v(x) = c_1 v_1(x) + c_2 v_2(x) + c_3 v_3(x) \text{ with } v_i(x) = N^1_{i-1}(x)$$

satisfies the boundary conditions; it consists of four segments and its derivative is constant for each segment.

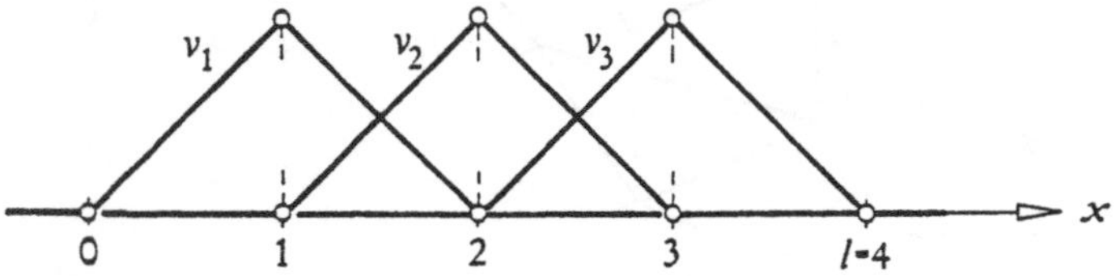

Figure 27.8 Linear B-splines as trial functions

The coefficients of the Ritz equations in **27.1** are composed of the terms

$$a_{i,j,k} = \int_{k-1}^{k} v'_{i,k}\, v'_{j,k} \; dx\,, \qquad a_{i,k} = \int_{k-1}^{k} f_k\, v_{i,k} \; dx\,.$$

The table below lists the value $v'_{i,k}$ etc. which are needed to determine these numbers:

		$k=1$	$k=2$	$k=3$	$k=4$
v'_{1k}	=	1	−1	·	·
v'_{2k}	=	·	1	−1	·
v'_{3k}	=	·	·	1	−1
$\int v_{1k}$	=	1/2	1/2	·	·
$\int v_{2k}$	=	·	1/2	1/2	·
$\int v_{3k}$	=	·	·	1/2	1/2
f_k	=	0	0	−0.4	−0.4

It is not difficult to find A and $\boldsymbol{a}$ for each individual segment from this table:

$$k=1 \qquad k=2 \qquad k=3 \qquad k=4$$

$$\begin{bmatrix}1 & \cdot & \cdot\\ \cdot & \cdot & \cdot\\ \cdot & \cdot & \cdot\end{bmatrix} \quad \begin{bmatrix}1 & -1 & \cdot\\ -1 & 1 & \cdot\\ \cdot & \cdot & \cdot\end{bmatrix} \quad \begin{bmatrix}\cdot & \cdot & \cdot\\ \cdot & 1 & -1\\ \cdot & -1 & 1\end{bmatrix} \quad \begin{bmatrix}\cdot & \cdot & \cdot\\ \cdot & \cdot & \cdot\\ \cdot & \cdot & 1\end{bmatrix}$$

$$0\cdot\begin{bmatrix}1/2\\ \cdot\\ \cdot\end{bmatrix} \quad 0\cdot\begin{bmatrix}1/2\\ 1/2\\ \cdot\end{bmatrix} \quad -0.4\cdot\begin{bmatrix}\cdot\\ 1/2\\ 1/2\end{bmatrix} \quad -0.4\cdot\begin{bmatrix}\cdot\\ \cdot\\ 1/2\end{bmatrix}.$$

The entries $a_{i,j,k}$ and $a_{i,k}$ which are zero because of what was remarked with regard to property (3) in **27.1** are marked by $\cdot$. The Ritz equations are now obtained by summation:

$$\begin{bmatrix}2 & -1 & \cdot\\ -1 & 2 & -1\\ \cdot & -1 & 2\end{bmatrix}\begin{bmatrix}c_1\\ c_2\\ c_3\end{bmatrix} - \begin{bmatrix}0\\ -0.2\\ -0.4\end{bmatrix} = \mathbf{o}\,.$$

They have the solution

$$c_1 = -0.2\,, \qquad c_2 = -0.4\,, \qquad c_3 = -0.4$$

which is depicted in Figure 27.9.

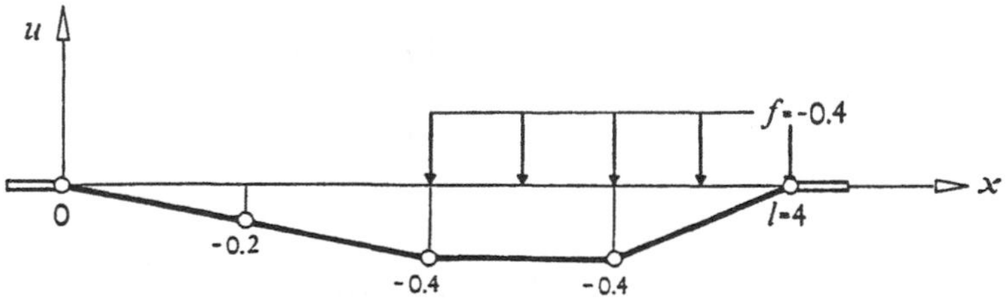

Figure 27.9 Loaded rope, linear spline

Example 2: The differential equation of Example 5 in **26.3** involves the second derivative. The trial functions for its numerical solution must therefore be at least quadratic. Assume that the rope has the length $l = 2$ and that it is loaded by $f = -1.6$ from $l/2$ to l. The trial function

$$v(x) = c_0 N^2_{-2}(x) + c_1 N^2_{-1}(x) + c_2 N^2_0(x) + c_3 N^2_1(x)$$

satisfies the boundary condition only if $c_0 + c_1 = 0$ and $c_2 + c_3 = 0$. This leads to the simpler function

$$v(x) = c_1 v_1(x) + c_2 v_2(x)$$

where

$$v_1(x) = N^2_{-1}(x) - N^2_{-2}(x)\,, \qquad v_2(x) = N^2_0(x) - N^2_1(x)\,.$$

It satisfies the boundary conditions, has only two segments and its second derivative is constant for each segment.

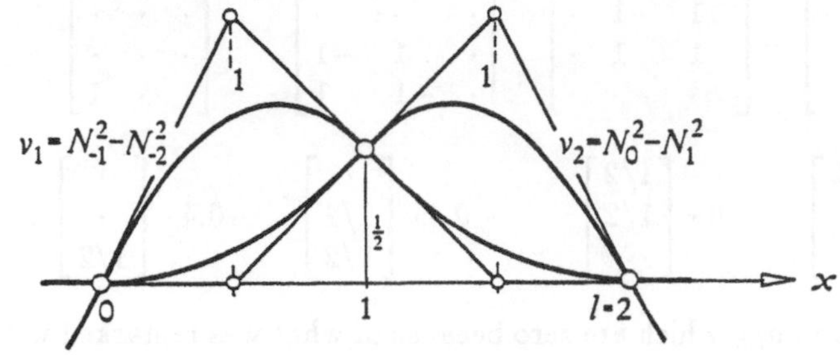

Figure 27.10 Quadratic splines as trial functions

The coefficients of the Galerkin equations in **27.3** are composed of the terms

$$a_{i,j,k} = \int_{k-1}^{k} (-v_{i,k} v''_{j,k}) dx\,, \qquad a_{i,k} = \int_{k-1}^{k} f_k \cdot v_{i,k}\, dx\,.$$

The table below lists the values $v_{i,k}$, etc., which are needed to find these numbers:

			$k=1$	$k=2$
Bézier ordinates	v_{1k}	:	$0, 1, 1/2$	$1/2, 0, 0$
	v_{2k}	:	$0, 0, 1/2$	$1/2, 1, 0$
	v''_{1k}	=	-3	1
	v''_{2k}	=	1	-3
	$\int v_{1k}$	=	$1/2$	$1/6$
	$\int v_{2k}$	=	$1/6$	$1/2$
	f_k	=	0	-1.6

The values $v''_{i,k}$ and $\int v_{i,k}$ can be determined directly from the Bézier points of $v_{i,k}$ as described in **20.2** and **22.6** respectively.

The matrix A and the vector a_k can easily be derived from the table. First one gets for each segment

$$k=1 \qquad\qquad\qquad\qquad k=2$$

$$\begin{bmatrix} 3/2 & -1/2 \\ 1/2 & -1/6 \end{bmatrix} \quad 0 \cdot \begin{bmatrix} 1/2 \\ 1/6 \end{bmatrix} \qquad\qquad \begin{bmatrix} -1/6 & 1/2 \\ -1/2 & 3/2 \end{bmatrix} \quad -1.6 \cdot \begin{bmatrix} 1/6 \\ 1/2 \end{bmatrix}$$

and on summing these parts the Galerkin equations

$$\begin{bmatrix} 4/3 & 0 \\ 0 & 4/3 \end{bmatrix} \begin{bmatrix} c_1 \\ c_2 \end{bmatrix} - \begin{bmatrix} -4/15 \\ -4/5 \end{bmatrix} = \mathbf{o}\,.$$

They have the solution

$$c_1 = -0.2\,, \qquad c_2 = -0.6\,,$$

which is depicted in Figure 27.11.

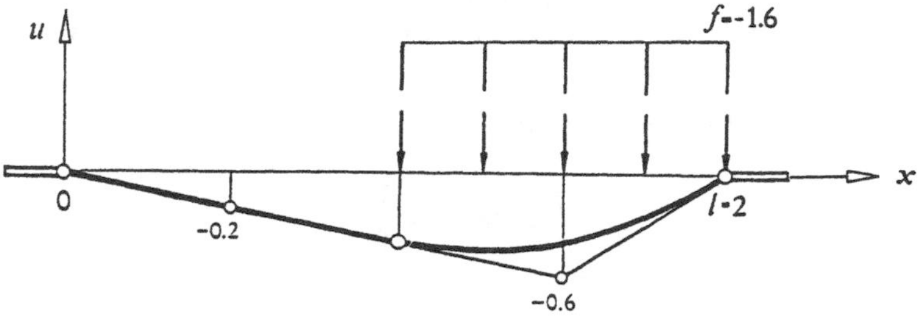

Figure 27.11 Loaded rope, quadratic spline

Example 3: Let $l = 4$ be the length of the diaphragm from **26.1** and $b = 1$ its width. Assume a uniform load $f = -0.4$ on the half from $l/2$ to l. Then

$$\frac{1}{2}\int_0^4\int_0^1 (u_x^2 + u_y^2)\,dy\,dx - \int_0^4\int_0^1 f \cdot u \; dy\,dx$$

describes the corresponding energy. Linear trial functions are sufficient to minimize the integral numerically. First, the domain G is triangulated, e.g., as in Figure 27.12.

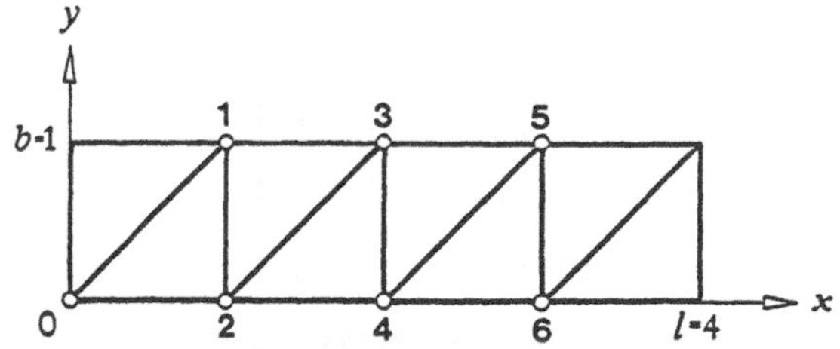

Figure 27.12 Triangulation of G

The linear combination

$$v(x,y) = c_1 v_1(x,y) + \cdots + c_6 v_6(x,y)$$

of the pyramids v_i with tips at the "knots" $1, \ldots, 6$ satisfies the boundary conditions and covers all eight subdomains. Its first partial derivatives are constant for each segment; they are given by the columns $[v_x, v_y]_i^t$ for the linear B-spline associated with the knot i in Figure 27.13.

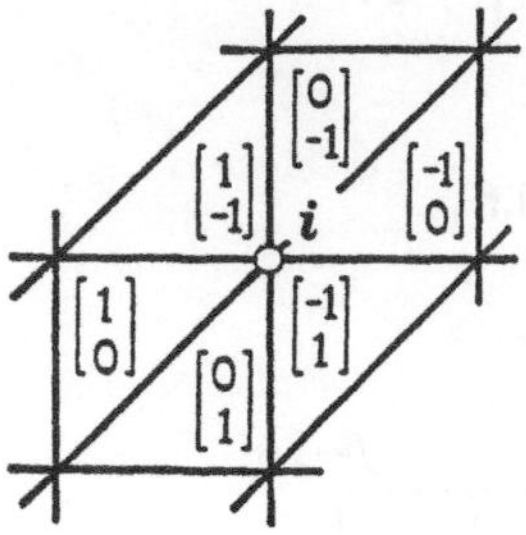

Figure 27.13
Linear B-spline with partial derivatives

For each triangle G_k of the support of v_i one has

$$\int_{G_k} 1 \; dx dy = \frac{1}{2}$$

and

$$\int_{G_k} v_i \; dx dy = \frac{1}{6}\,.$$

From this one readily establishes the coefficients

$$a_{i,j} = \int_G [v_x v_y]_i \begin{bmatrix} v_x \\ v_y \end{bmatrix}_j , \qquad a_i = \int_G f \cdot v_i$$

of the Ritz equations, namely

$$\begin{bmatrix} 2 & -1 & -\frac{1}{2} & \cdot & \cdot & \cdot \\ -1 & 2 & 0 & -\frac{1}{2} & \cdot & \cdot \\ -\frac{1}{2} & 0 & 2 & -1 & -\frac{1}{2} & \cdot \\ \cdot & -\frac{1}{2} & -1 & 2 & 0 & -\frac{1}{2} \\ \cdot & \cdot & -\frac{1}{2} & 0 & 2 & -1 \\ \cdot & \cdot & \cdot & -\frac{1}{2} & -1 & 2 \end{bmatrix} \begin{bmatrix} c_1 \\ c_2 \\ c_3 \\ c_4 \\ c_5 \\ c_6 \end{bmatrix} - \begin{bmatrix} 0 \\ 0 \\ -\frac{1}{15} \\ -\frac{2}{15} \\ -\frac{1}{5} \\ -\frac{1}{5} \end{bmatrix} = 0\,.$$

They have the solution

$$c_1 = c_2 = -0.2 \pm \eta\,, \qquad c_3 = c_4 = -0.4 \pm 6\eta\,, \qquad c_5 = c_6 = -0.4 \pm \eta$$

where

$$\eta = 0.002\,.$$

It is depicted in Figure 27.14.

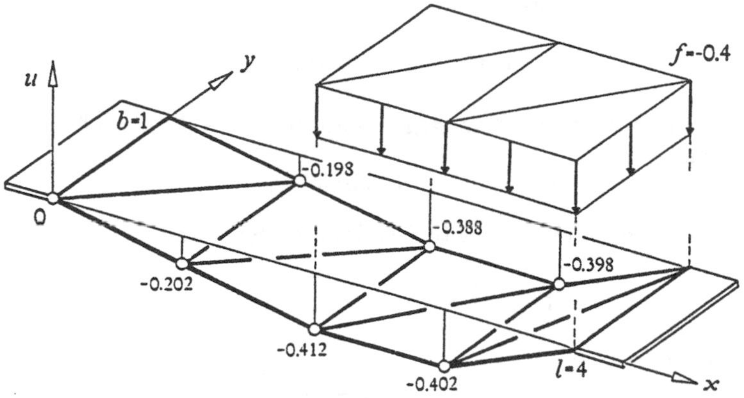

Figure 27.14 Loaded diaphragm, "method of plates"

27.5 Local Coordinates

Local coordinates were already used in **21.1** to describe the individual segments. They were also used — but it was not explicitly mentioned — in the examples of **27.4**. The use of such local coordinates is an important aid in the finite element method to describe the segments. They facilitate standardizations and thus significantly simpler employment of the method.

The easiest local coordinates are obtained by linear interpolation **18.5** on a line $\boldsymbol{x}_0, \boldsymbol{x}_1$,

$$\boldsymbol{x} = \boldsymbol{x}_0(1-\xi) + \boldsymbol{x}_1\xi\,, \qquad \xi \in [0,1]\,,$$

and bilinear interpolation **18.11** on a triangle $\boldsymbol{x}_{0,0}$, $\boldsymbol{x}_{1,0}$, $\boldsymbol{x}_{0,1}$, or a quadrilateral $\boldsymbol{x}_{0,0}$, $\boldsymbol{x}_{1,0}$, $\boldsymbol{x}_{0,1}$, $\boldsymbol{x}_{1,1}$,

$$\boldsymbol{x} = [\boldsymbol{x}_{0,0}(1-\xi) + \boldsymbol{x}_{1,0}\xi](1-\eta) + [\boldsymbol{x}_{0,1}(1-\xi) + \boldsymbol{x}_{1,1}\xi]\eta\,, \qquad \xi,\eta \in [0,1]\,.$$

If the $\boldsymbol{x}_{i,j}$ form a parallelogram as in Figure 27.15, i.e., $\boldsymbol{x}_{0,0} - \boldsymbol{x}_{1,0} - \boldsymbol{x}_{0,1} + \boldsymbol{x}_{1,1} = \boldsymbol{0}$, then this can be rewritten as

$$\boldsymbol{x} = \boldsymbol{x}_{0,0} + [\boldsymbol{x}_{1,0} - \boldsymbol{x}_{0,0}]\xi + [\boldsymbol{x}_{0,1} - \boldsymbol{x}_{0,0}]\eta\,.$$

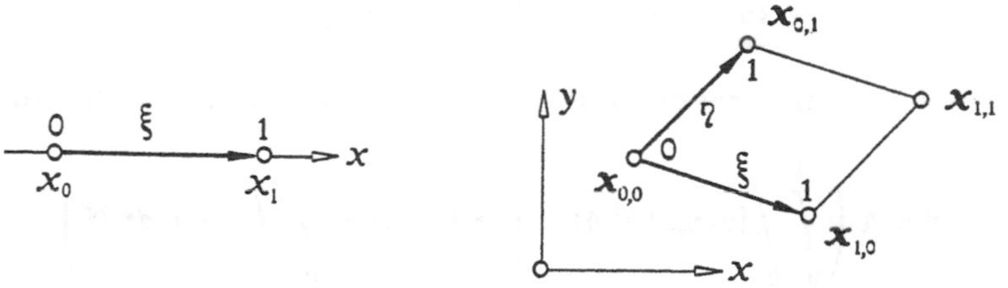

Figure 27.15 Local coordinates; linear and bilinear interpolation

Along with a coordinate change goes an alteration of the energy integrals of **26.2** and **26.3**. Differentiation yields

$$dx = a\ d\xi \quad \text{and} \quad d\mathbf{x} = A\ d\boldsymbol{\xi}$$

where $a = x_1 - x_0$, $A = [\mathbf{x}_{1,0} - \mathbf{x}_{0,0}, \mathbf{x}_{0,1} - \mathbf{x}_{0,0}]$ and $\boldsymbol{\xi} = [\xi, \eta]^t$. Then, on using the chain rule one gets

$$\dot{u} = au', \qquad \ddot{u} = a^2 u'', \ldots$$

(the ordinary derivative with respect to ξ is denoted by $\cdot$) and

$$[u_\xi, u_\eta] = [u_x, u_y]A, \text{ etc.}$$

This can be reversed

$$u' = \frac{1}{a}\dot{u}, \qquad u'' = \frac{1}{a^2}\ddot{u}, \ldots,$$

$$[u_x, u_y] = [u_\xi, u_\eta]A^{-1}$$

and substituted into the energy integrals. Furthermore, the integration elements dx and $dxdy$ need replacement by

$$a\ d\xi \quad \text{and} \quad \det A\ d\xi d\eta \quad \text{respectively.}$$

The domains of integration become simpler after local coordinates have been introduced in the subdomains G_k thus making it possible to evaluate the integrals over all G_k in a uniform fashion.

Remark 4: The coefficients of the quadratic form q, of the linear form l, or of the differential operator $\mathcal{L}$ are not always constant and this makes the integration more difficult. In general it suffices to approximate the coefficients over the subdomains G_k by constant or linear (bilinear) splines. The same can be said about the function f, cf. Figure 27.14.

Example 4: Consider the integral of Example 1 in **26.1**. One gets

$$\frac{1}{2a}\int_0^1 (\dot{u})^2 d\xi - a\int_0^1 f \cdot u\ d\xi$$

for the contribution to the energy over the interval $[x_0, x_1]$.

Example 5: For the integral of Example 2 in **26.1** one has the contribution

$$\det A \left(\int_0^1\int_0^1 [u_\xi u_\eta][A^t A]^{-1} \begin{bmatrix} u_\xi \\ u_\eta \end{bmatrix} dy d\xi - \int_0^1\int_0^1 f \cdot u\ d\eta d\xi \right)$$

to the energy over a parallelogram $\mathbf{x}_{0,0}, \mathbf{x}_{1,0}, \mathbf{x}_{0,1}, \mathbf{x}_{1,1}$.

Example 6: Let the interval [0,4] of Example 1 in **27.4** be divided into two subintervals only. On introducing the local coordinates

$$x = 2\xi_1 \quad \text{and} \quad x = 2\xi_2 + 2$$

one obtains

$$\int_0^4 \left(-\frac{1}{2}u'' - f\right) u \; dx = \frac{1}{2}\left(\int_0^1 \left(-\frac{1}{2}\ddot{u} - 4f\right) u \; d\xi_1 + \int_0^1 \left(-\frac{1}{2}\ddot{u} - 4f\right) u \; d\xi_2\right)$$

for the energy integral in **26.3** which is to be minimized. This problem was already solved in this form in Example 2 of **27.4**.

27.6 Notes and Exercises

1. Originally the finite element method was mainly used to compute the nodal displacement in frameworks. Therefore, the variables c_i are sometimes called **node variables**.
2. Trial functions which satisfy the physical and mathematical continuity requirements are called **conformal**. Non-conformal trial functions are less expensive to compute, and, frequently, yield useful results in practice.
3. Trivariate trial functions over three-dimensional domains G can be defined in analogy to uni- and bi-variate trial functions. The domain G is best subdivided into tetrahedra, cuboid- or prism like subdomains G_k.
4. Often it suffices to approximate domains G with curved boundaries by subdomains G_k with straight boundaries. When higher accuracy is desired, one approximates the boundary by quadratic splines. Then, the curved elements G_k are called **isoparametric** if the transformation to local coordinates is of the same form as the trial functions associated with this element.

28 Bibliography

There are numerous textbooks on Numerical Analysis. Below are some selected references which are listed for a further reading in this subject or because of their different point of view.

Bulirsch, R. and *J. Stoer*: Introduction to Numerical Analysis, Springer, New York 1980

Conte, S. D. and *C. de Boor*: Elementary Numerical Analysis, McGraw-Hill, New York 1980

Dahlquist, G. and *A. Björck*: Numerical Methods, Prentice-Hall, Englewood Cliffs 1974

Henrici, P.: Elements of Numerical Analysis, J. Wiley & Sons, New York 1964

Kohn, M. C.: Practical Numerical Methods: Algorithms and Programs, Macmillan, New York 1987

Marchuk, G. I.: Methods of Numerical Mathematics, Springer, Berlin 1982

Schwarz, H. R. and *J. Waldvogel*: Numerical Analysis: A comprehensive introduction, J. Wiley & Sons 1989

Stiefel, E.: An Introduction to Numerical Analysis, Academic Press, New York 1963

Special topics of Numerical Analysis are considered, e.g., in:

Alefeld, G. and *J. Herzberger*: Introduction to Interval Computations, Academic Press, London 1983

Boehm, W., G. Farin and *J. Kahmann*: A survey of Curve and Surface Methods in CAGD, Computer Aided Geometric Design 1 (1984), 1-60

de Boor, C.: A Practical Guide to Splines, Springer, New York 1968

Collatz, L.: Functional Analysis and Numerical Mathematics, Academic Press, London 1966

Davis, P. J.: Interpolation and Approximation, Blaisdell, New York 1963

Prenter, P. M.: Splines and Variational Methods, J. Wiley & Sons, New York 1975

Schwarz, H. R.: Numerical Analysis of Symmetric Matrices, Prentice-Hall, Englewood Cliffs 1973

Wilkinson, J. H.: The Algebraic Eigenvalue Problem, Oxford University Press, London 1965

Wilkinson, J. H. and *C. Reinsch*: Linear Algebra, Handbook for Automatic Computation, Vol II, Springer, Berlin 1971

Zienkiewicz, O.: The Finite-Element Method in Engineering Science, McGraw-Hill, London 1977

BASIC programs to most of the algorithms presented in this book can be found in:

Kahmann, J.: BASIC-Programme zur Numerischen Mathematik, Vieweg, Braunschweig 1984

Index

Page numbers in *italics* refer to algorithms.

Printed and bound by CPI Group (UK) Ltd, Croydon, CR0 4YY
23/10/2024
01777695-0004